HANDBOOK OF GEOSTATIONARY ORBITS

Handbook of Geostationary Orbits

by

E. M. Soop

European Space Operations Centre

KLUWER ACADEMIC PUBLISHERS
DORDRECHT / BOSTON / LONDON

MICROCOSM, INC.
TORRANCE, CALIFORNIA

Library of Congress Cataloging-in-Publication Data

Soop, E. M.
   Handbook of geostationary orbits / by E.M.Soop.
      p.   cm. -- (Space technology library ; v. 3)
   Includes bibliographical references and index.

   1. Geostationary satellites. 2. Artificial satellites--Control
systems--Data processing. 3. Artificial satellites--Orbits.
4. PEPSOC.   I. Title.   II. Series.
TL796.6.E2S66   1994
629.4'113--dc20   94-29408
                                                     - - -

ISBN 978-90-481-4453-2

Published jointly by Microcosm, Inc.,
2601 Airport Drive, Suite 230,
Torrance, California 90505, U.S.A.
and Kluwer Academic Publishers,
P.O. Box 17, 3300 AA Dordrecht, The Netherlands.

Kluwer Academic Publishers incorporates
the publishing programmes of
D. Reidel, Martinus Nijhoff, Dr W. Junk and MTP Press.

Sold and distributed in the U.S.A. and Canada
by Kluwer Academic Publishers,
101 Philip Drive, Norwell, MA 02061, U.S.A.

In all other countries, sold and distributed
by Kluwer Academic Publishers Group,
P.O. Box 322, 3300 AH Dordrecht, The Netherlands.

Printed on acid-free paper

CONTENTS

Foreword

This *Handbook of Geostationary Orbits* is in principle an extension of the *Introduction to Geostationary Orbits* that was printed as a special publication by the European Space Agency (ESA) in 1983. The immediate purpose was to provide the theoretical background and some practical advice for the orbit control of geostationary spacecraft by means of the software package "PEPSOC".

PEPSOC, short for "Portable ESOC Package for Synchronous Orbit Control", was produced by the European Space Operations Centre (ESOC) to support spacecraft operations in the routine phase. The resulting publication was a handbook for engineers and spacecraft operators, rather than a classical textbook in celestial mechanics.

During the past eleven years, the software system PEPSOC has found a wide application both within and outside the ESA member states. At the same time, the original *Introduction* found numerous readers also outside the group of PEPSOC operators. The continuing development and the increasing use of the geostationary orbit has now created the need for a new, more detailed publication to include new aspects that have emerged.

The present *Handbook* contains several additional subjects and more mathematics to describe the methods applied in PEPSOC. The geophysical and astronomical parameters have been updated to reflect the latest recommended values. This results in small deviations of the numerical data compared to the *Introduction*.

Other changes concern the definition of the mean longitude drift rate, denoted by D, which now has become a dimension-less fraction of the Earth's rotation rate. The four coefficients that are defined in Chapter 8 to describe the time variation of the differentials of the tracking measurements were previously called C_1, C_2, C_3, C_4, but are now denoted by b,u,v,w with a slightly different definition.

New subjects are, in particular:

- Legal aspects, in Section 1.3.

- Re-orbiting of old spacecraft, in Section 3.4.

- Algorithm for the gradient of spherical harmonics, in Section 3.4.

- Eclipse by Moon, in Section 5.4.

- Co-location, in Sections 5.5, 5.6 and 5.7.

- Long-term inclination strategy, in Section 6.4.

- Numerical method for longitude manoeuvres, in Section 7.6.

- Light-time equation for tracking, in Section 8.2.

- Statistics of tracking errors, in Section 8.6.

- Orbit accuracy estimates, in Section 8.7.

The author wishes to thank all readers whose comments and questions have inspired the extension that led to the present *Handbook*. The preparation of the camera-ready copy was made possible by the invaluable assistance of many ESOC colleagues, in particular by Fabienne Delhaise, who produced the new figures by means of a graphical software system.

Darmstadt, in April 1994
Erik Mattias Soop

1. GENERAL BACKGROUND

1.1 Introduction

The number of spacecraft in geostationary orbit is increasing steadily, from the first one launched in 1963 to more than one hundred in 1980 and two hundred in 1990. There is no indication yet of any saturation of interest from the user community. This type of orbit, in which the spacecraft is nearly at rest with respect to the rotating Earth, is used mainly by communications missions, but also by some Earth observation missions and a few scientific missions. The main advantage is the permanent contact between the ground station and a spacecraft in a geostationary orbit and, for Earth observations, the possibility of constantly surveying the same geographical region.

The strictest requirements on the geostationary properties of the orbit arise with communications missions, where the ground stations are equipped with fixed-direction antennas. At the other end of the scale, some scientific spacecraft do not require any other benefit from the geostationary orbit than continuous visibility from one ground station with an autotrack antenna. The radio regulations, however, impose constraints on the longitude variations of the spacecraft in order to minimise the frequency interference with neighbours in orbit.

We will use the expression *geostationary* to characterise a mission where one is aiming to keep the spacecraft as far as possible at rest, relative to the Earth. The word "geosynchronous" is often used in the same sense, although it actually only means that the orbital period coincides with the rotation of the Earth, but it does not impose any restrictions on the orbital eccentricity or inclination.

The scope of this text is in principle limited to a description of the geostationary part of a mission so only a very brief outline of the launch and early operations phase (LEOP) activities will be given here. A few more details on the LEOP operations are given in Section 6.2 in connection with the achievement of the initial orbital inclination. The data given in the following refer to the launcher Ariane, but LEOP operations with other launchers are performed in a similar manner.

After 15 minutes of powered flight, the launcher puts the spacecraft at about 200 km height above the Earth into a transfer orbit (TO) with its apogee near the geostationary height, Figure A. After a few revolutions in the transfer orbit, the apogee motor is fired (AMF) to inject the spacecraft into the geostationary orbit. The AMF fuel makes up almost half the spacecraft weight and is used to increase the flight velocity from 1.6 km/s to 3 km/s. At the same time the flight direction is changed so that the transfer orbit inclination of around 7° is decreased to the near 0° geostationary inclination.

Early geostationary spacecraft had apogee motors with solid fuel that burnt out in less than one minute. More recent designs use liquid fuel in a motor that can be started and stopped several times at different apogees. Between the firings the spacecraft flies in intermediate orbits with higher perigees but the apogee still near the geostationary position. The velocity increments of the different firings add up to the same value as the single thrust shown in Figure A.

After AMF the station acquisition starts, usually with a spin axis erection or a three-axis acquisition. A series of smaller orbit manoeuvres is performed over a period of up to one month to move the spacecraft to the desired longitude and to adjust the orbit eccentricity and inclination.

The station acquisition orbit manoeuvres compensate for errors in the AMF burn direction, time and magnitude, in addition to possibly remaining errors from the launcher injection. There are usually more orbit parameters to be adjusted after a solid fuel than a liquid fuel AMF. In the former case, the burn size is fixed, so only three free parameters of AMF can be controlled from ground: The two direction angles and the time of firing.

When the station acquisition is completed the routine operations can start. This phase of the mission usually lasts several years until the on-board fuel is exhausted, the electric power generators have deteriorated, a major on-board error has occurred or the mission becomes obsolete.

After such an end of the mission the spacecraft would remain for ever drifting around in and near the geostationary region. In later years, the potential problem of the increasing accretion of abandoned geostationary spacecraft has attracted the attention of leading space agencies. It is now recommended that, at the end of the mission, old geostationary spacecraft are moved into a circular orbit a few hundred km above the geostationary altitude by several tangential east thrusts, as described in Section 3.4.

So far there has been no reported collision in space between two operational geostationary spacecraft. As the number of spacecraft increases, however, there is a growing risk of damage to, in particular, the very large solar panels that are used on recent missions.

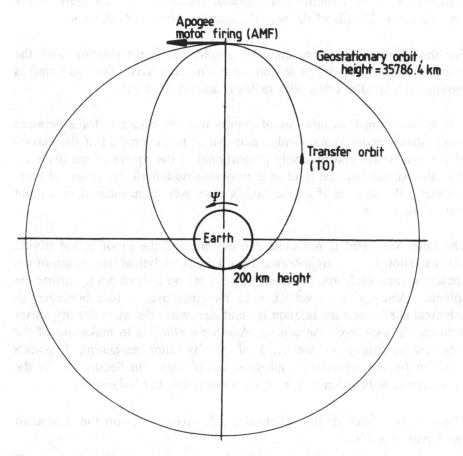

Figure 1.1.A. The launch and early orbit phase (LEOP): Transfer orbit, apogee motor firing and the geostationary orbit. The orbits and the Earth are drawn to scale, but in reality the transfer orbit is not in the same plane as the geostationary orbit.

1.2 The Geostationary Orbit

A perfectly geostationary orbit is a mathematical abstraction that could be achieved only by a spacecraft orbiting around a perfectly symmetric Earth with no other forces acting on the spacecraft than the central gravity attraction from the Earth. The abstraction is, however, useful as an approximate description of the real case, since all other forces, from the Moon and the Sun and from the nonspherical part of the Earth's gravity, are small in comparison. It is customary in celestial mechanics to call these forces *perturbations*. Details of the perturbations are given in Chapter 4.

For the idealised situation, consider a spherical Earth rotating with the constant angular velocity ψ around the north-south axis. The spacecraft is moving in a circular orbit, with radius r, around the Earth.

It is known from Newton's law of gravity that the attractive force between two bodies is proportional (with a constant g) to the product of the masses of the two bodies and inversely proportional to the square of the distance. It is also known that the field of gravity outside a perfectly spherical body is exactly the same as if all the body's mass were concentrated as a point mass at its centre.

The latter statement is not quite self-evident and the proof is not trivial. One can show it by straight-forward but tedious analytical integration of the attraction from each mass element of a sphere on a fixed point outside the sphere. Actually, it is sufficient to integrate over a thin homogenous spherical shell, since the relation is valid also when the mass density varies with the distance from the centre. Another method is to make use of the spherical symmetry of the field of gravity after separating Laplace's equation for the potential in spherical coordinates. In Section 4.2 is the gravitational field around the real, unsymmetrical, Earth described.

The attractive force of the spherical Earth (mass = M) on the spacecraft (with mass = m) is:

$$F = gmM/r^2$$

It is customary to express this by means of the Earth's central gravity constant $\mu = gM$ since it is known with a much higher accuracy than g and M separately.

The centrifugal force of the spacecraft's motion in the orbit must balance the attraction force. By adjusting the orbital radius we can obtain a circular orbit with the same angular velocity as the Earth's rotation:

$$m\psi^2 r = m\mu/r^2$$

One can divide both sides of the above equation by m so the equation of motion becomes independent of the spacecraft mass. The balance is possible for only one value of r, namely:

$$r = \sqrt[3]{\mu/\psi^2}$$

The Earth's rotation rate ψ is known with a very high accuracy as follows:

$$\psi = 360.985647 \text{ deg/day} = 0.729211585 \times 10^{-4} \text{ rad/s}$$

For the Earth's gravity, the value adopted in 1989 by the International Earth Rotation Service is

$$\mu = 398600.440 \text{ km}^3/\text{s}^2$$

Inserting these numbers we obtain r = 42164.2 km. However, when all other forces are taken into account, as described in Section 4.3, one obtains the mean value of the actual

geostationary radius = 42164.5 km

It varies slightly in time because of the time-dependent perturbations on the orbit. An earlier value of μ that was previously recommended by the International Astronomical Union was about 1 km³/s² higher, which made the geostationary radius 30 m higher.

A further criterion on the orbit, for being geostationary, is to lie in the equatorial plane with the spacecraft motion in the same direction as the rotation of the Earth, i.e. eastward, as shown in Figure A. With these idealised assumptions, the spacecraft is at rest relative to the Earth. It must be stationed above the equator, but the *subsatellite longitude*, i.e. the longitude of the projection of the spacecraft on the Earth's surface, can be selected arbitrarily. In fact, the longitude is the only free parameter that is available when geostationary orbits are allocated to different space missions.

In practice, of course, the spacecraft will not stay indefinitely in the same position relative to the Earth because additional forces acting on it will

change the shape of the orbit, the orientation of the orbital plane and the
spacecraft longitude. These changes can be compensated for by active orbit
control, i.e. orbit manoeuvres performed by activating thrusters on-board the
spacecraft, usually by manual ground control. These are called *station
keeping* manoeuvres.

A truly geostationary orbit can only exist instantaneously, since the velocity
of the spacecraft relative to the Earth can be zero only for an instant. In
practice one obtains an approximately geostationary orbit by minimising the
motion relative to the Earth by means of the station keeping manoeuvres.
The time interval between the manoeuvres determines how close the actual
orbit is to the ideal geostationary orbit.

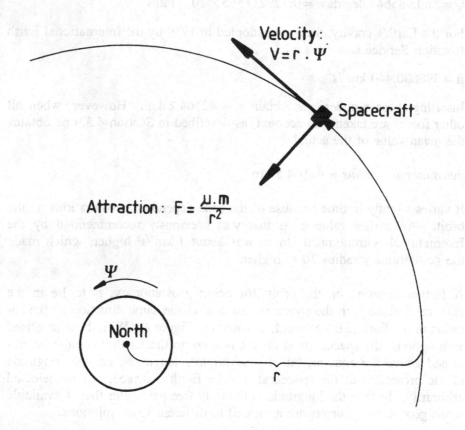

Figure 1.2.A. The geostationary orbit seen from the north.

For the mean radius of the Earth at the equator we use here the value of 6378.144 km, but other values that deviate by a few metres are also in use. The spacecraft height above the Earth's surface is 35786.4 km, corresponding to 5.61 Earth radii. At this height, the gravitational attraction from the Earth is 0.2242 m/s². The spacecraft velocity in an inertial co-ordinate system equals:

$$V = \psi r = 3.075 \text{ km/s}$$

1.3 Legal Aspects

It was shown in the preceding section that the geostationary orbit constitutes a very limited region of space. This region is defined by the distance from the Earth, which must be near the geostationary radius, and the latitude relative to the Earth's equator, which must be near zero. The ring-shaped region that remains has only *one* free dimension to allocate to different spacecraft, namely the longitude relative to the Earth (subsatellite longitude).

The history of utilisation of subsatellite longitudes by geostationary spacecraft followed the same evolution as the exploitation of other natural resources: The first few users took whatever they wanted, and a certain amount of coordination of the usage was introduced only when it was found to be necessary. For historical and practical reasons this was combined with the allocation of radio frequencies, since most geostationary spacecraft are communications spacecraft and most communications spacecraft are in geostationary orbit.

The allocation of frequencies for terrestrial radio communications was first coordinated in 1906 as a continuation of international agreements about telegraph and telephone transmissions. The administration was taken over in 1947 by the International Telecommunication Union (ITU), which was established as a specialised agency of the United Nations. All decisions are taken by the World Administrative Radio Conferences (WARC), which are organised by ITU every few years. Only sovereign states can send delegates to WARC, and international organisations are represented indirectly through their member states. Most organisations, like the European Space Agency (ESA) also sends observers to WARC.

The first space age WARC in 1959 was extended to include radio communications to and from space vehicles, in addition to the traditional terrestrial communications. The 1971 WARC recognised the geostationary orbit to be a "limited natural resource" like the radio frequency spectrum. In 1973 the allocation of longitude positions on an "equitable access" basis was added to the responsibilities of WARC, to be administered by ITU. The 1977 WARC began the task of allocating and assigning longitude slots. In many cases, the same longitude position was given to several spacecraft belonging to both the same and to different owners.

Many nations, also without access to space technology, requested longitude positions for possible future use because of fear of losing access to this important resource. A claim in 1976 by a group of equatorial nations for sovereignty over the geostationary longitudes above their territory did not obtain a positive response by the spacefaring states.

At the beginning, the allocation of longitude positions was performed only for the purpose of avoiding signal interference between neighbouring spacecraft that utilise the same radio frequency, whereas the risk of collision was estimated to be negligible. This was justified by the consideration that a typical shared deadband is of the order of more than 100 km wide in longitude and latitude and about half as deep in the radial direction. Only later did some space agencies realise that the potential risk of physical collision between them is not negligible. However, so far no collision has been reported between two spacecraft in operation, which is the probable reason why WARC has not yet been asked to take action.

Figure A shows the longitude distribution along the geostationary longitudes of 481 operational spacecraft and other objects, like dead spacecraft and burnt-out rocket stages etc. in mid-1993.

The attitude towards the collision risk differs strongly between different space agencies. Some neglect the risk and others spend a great deal of effort to minimise it. When the co-located spacecraft are operated by the same control centre it becomes relatively straightforward to plan the manoeuvres so as to avoid close encounters, as shown in Sections 5.5, 5.6 and 5.7. With different control centres a safe co-location becomes more cumbersome, but there are sometimes bilateral arrangements made for avoiding any collision risk. However, there are no generally agreed traffic rules for space flight like for road traffic, and there is no general law concerning the liability for damage by collision.

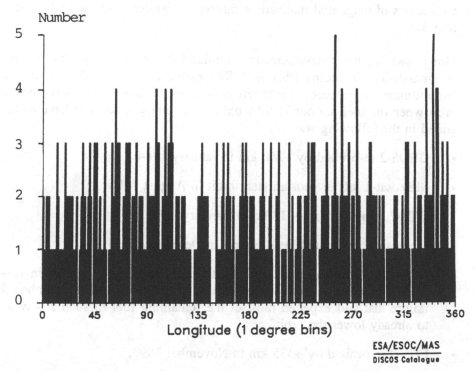

Figure 1.3.A. Number of objects in geostationary and near-geostationary orbit per 1° longitude slot in mid-1993. The total number of objects = 481, of which 286 are spacecraft in operation. Courtesy of the Mission Analysis Section, ESOC.

ESA's Olympus communications spacecraft was operated from mid-1989 to mid-1990 in the longitude slot 19.0°W ± 0.07° in co-location with one German and one, later two, French spacecraft. The inter-spacecraft distances were checked and the manoeuvres coordinated in an ad hoc manner for one year, after which the other spacecraft were shifted by 0.2° east and west, respectively, and Olympus was left alone at 19.0°W.

There is a real risk that the future usage of the geostationary orbit will be hampered by the accumulation of old, abandoned spacecraft. In order to protect this "limited natural resource", the United Nations Committee on the Peaceful Uses of Outer Space has recommended all space centres to remove ("re-orbit") old geostationary spacecraft before their end of service. In most cases the orbit is raised by a few 100 km above the geostationary radius by a series of tangential manoeuvre thrusts, as explained at the end of Section 3.4.

Many space agencies have already committed themselves to follow this recommendation, including ESA in 1989. However, before ESA made this commitment, one spacecraft was left to drift uncontrolled in a geostationary orbit when the fuel ran out. ESA's old geostationary spacecraft have so far ended in the following way:

- GEOS-2 re-orbited by +260 km in January 1984.

- Meteosat-1 left in geostationary orbit in August 1985.

- OTS-2 re-orbited by +318 km in January 1991.

- Meteosat-2 re-orbited by +334 km in December 1991.

- Olympus re-orbited *below* the geostationary orbit, by -213 km in August 1993 by ground commanded thrusts following upon an onboard failure that had expelled most of the remaining fuel in such a way as to already lower the orbit.

- ECS-2 re-orbited by +335 km in November 1993.

2. FUNDAMENTAL DEFINITIONS

2.1 Co-ordinate Systems

The need for well-defined co-ordinate systems for expressing the instantaneous position of a spacecraft comes from several sources:

- In order to ensure that the spacecraft satisfies the geostationary requirements one must know its position relative to the Earth;

- The tracking measurements for the orbit determination define the spacecraft position relative to the ground stations;

- The equations of motion that describe the spacecraft flight are valid only in an inertial system;

- The positions of the Sun and the Moon must be modelled in a known co-ordinate system so that their attraction on the spacecraft can be taken into consideration.

All these requirements cannot be satisfied by a single system, so we must introduce a series of co-ordinate systems, connected by well-defined time-dependent translations and rotations. For geostationary spacecraft, however, the situation is easier than for many other types of missions. Since the geostationary position is defined in the same system in which the tracking is performed one can use a simpler approximation for the transformations than for other orbits. For this reason only a brief summary of the transformations will be given here.

An inertial system is needed in which Newton's equation of acceleration is valid: the acceleration of a mass point with position vector denoted by $\bar{r}$ equals the force divided by the mass:

$$\frac{d^2\bar{r}}{dt^2} = \frac{\bar{F}}{m}$$

For the equation of motion we do not need more advanced physics, like the Theory of Relativity, but some influence of it appears in connection with solar radiation pressure (Section 4.5) and the light-time equation (Section 8.2).

From astronomical measurements it has been found that a suitable inertial system can take the mass centre of our planetary system as reference point and the directions to the fixed stars as reference directions. In this system, the Sun and the planets move according to Newton's equation under mutual gravitational attraction. We will use this system only for reference purposes, but for practical calculations of orbits around the Earth it is easier to locate the origin of the co-ordinate system at the mass centre of the Earth.

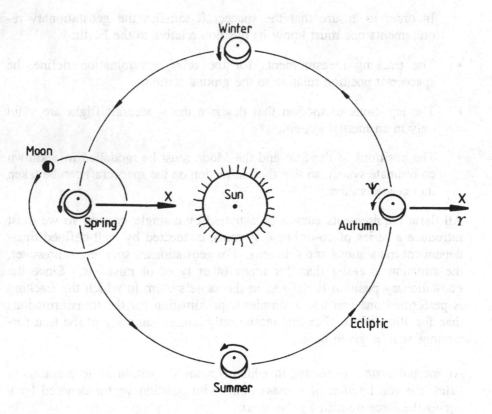

Figure 2.1.A. Schematic diagram, not to scale, of the Earth's orbit around the Sun (=ecliptic), seen from north. The Moon's orbit is inclined 5.14° to the ecliptic plane. The direction of the vernal equinox (x-axis) is to the right.

The orbit of the Earth around the Sun is called the *ecliptic*. It is nearly circular with a distance that varies from 147 to 152 million km and with the period of 1 year = 365.2422 days, Figure A. The standard system for the geostationary orbit will then be the quasi-inertial *Mean Equatorial Geocentric System of Date* "MEGSD". In order to use Newton's equation of motion for the spacecraft in this co-ordinate system we put the contributions from the acceleration of the system at the right-hand side of the differential equation together with the other perturbations. The same is done with the contributions from the *precession* (small rotation) of the system that is further described below.

The x-y-plane of our co-ordinate system is taken as the equatorial plane of the Earth. The Earth rotates around the z-axis in a mathematically positive direction, Figures A and B. The x-direction is chosen to coincide with the intersection of the equatorial plane and the plane of the ecliptic. This direction is known in astronomy as the *Vernal Equinox* or the *First Point of Aries* and is marked with the Aries head that looks like ♈. In this system the Sun and the Moon are seen to rotate in the positive direction around the Earth.

Unfortunately, the use of the MEGSD co-ordinate system is complicated by the fact that both the equatorial plane and the ecliptic move slowly with respect to the inertial system. The details of the transformations are given in publications from the International Astronomical Union, the International Earth Rotation Service, the Royal Greenwich Observatory and in the Explanatory Supplement to the Astronomical Ephemeris. The actual rotational motion of the Earth in an inertially oriented reference frame is by internationally adopted convention split into the following parts:

- precession
- nutation
- uniform rotation
- rotation with time correction
- polar motion

The *precession* of the Earth is the secular effect of the gravitational attraction from the Sun and the planets on the equatorial bulge of the Earth. The main effect is a rotation of the mean-of-date system in the negative sense in the ecliptic plane by one turn in 26000 years, which is equivalent to 0.014° per year.

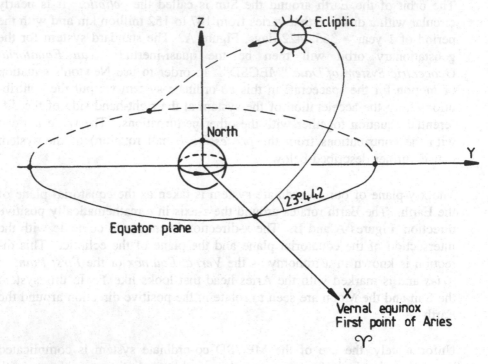

Figure 2.1.B. The geocentric equatorial co-ordinate system.

The *nutation* is the short-periodic effect of the gravitational attraction of the Moon and, to a lesser degree, the planets on the Earth's equatorial bulge. It has a certain periodicity with important contributions from the Moon's orbital period. The maximum value of any one of the two nutation angles is 0.006°.

The *polar motion* is the motion of the Earth's pole (= the axis of rotation) relative to the fixed Earth, as opposed to the nutation which defines its motion in MEGSD. The pole moves in a path that resembles a spiral with a period of slightly less than a year with a size of the order of 20 metres on the surface of the Earth.

The *Mean of Date* in the name of MEGSD means that it is transformed from the inertial system only by the precession but uses the mean value of zero for the other more or less periodic transformations. Different systems are often used, however, by different space operations centres. The advantage of using the MEGSD co-ordinate system is that for all purposes except high-precision orbit determination we can consider the Earth to rotate with a uniform angular velocity around the z-axis in the system. This rotation is the only co-ordinate transformation that is needed for the understanding of the flight of geostationary spacecraft in what follows. Positions on the Earth are defined by longitude and latitude where the longitude is referred to the zero meridian or the Greenwich meridian, originally drawn through the Royal Greenwich Observatory in England.

The time scale used for all satellite operations is called UTC (Co-ordinated Universal Time, previously Greenwich Mean Time), by which also all satellite telemetry and tracking data is labelled at the ground stations. UTC is piecewise uniform and continuous except when *leap-seconds* are inserted. This is scheduled so that it preferably takes place at 0 hours UTC on January 1 or July 1. The UTC clocks are then stopped for one second in order to allow the Earth's rotation to catch up, as explained below.

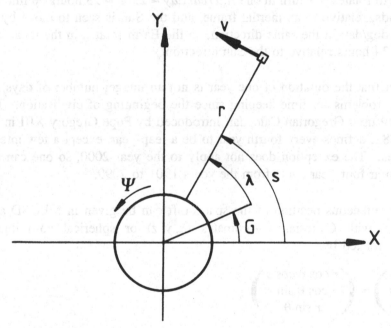

Figure 2.1.C. Sidereal angles of Greenwich (*G*) and spacecraft (*s*) and the spacecraft's longitude ($\lambda = s - G$) seen from north.

The Greenwich sidereal angle, which is the angle from the x-axis to the zero meridian (Figure C), can be calculated as a function of UTC time with $0.004°$ accuracy by means of the uniform angular velocity that we will denote by ψ in the following.

$$G = G_0 + \psi(t - t_0)$$

$$\psi = 0.729211585 \times 10^{-4} \text{ rad/s} = 360.985647 \text{ deg/day}$$

Here G_0 is the value of G at an epoch t_0. Table 1 shows the values of G_0 at 0 hours UTC on January 1 for a sequence of years, derived from *Newcomb's formula*. For higher accuracy one must include more terms and also the *time correction* to model small deviations in the Earth's rotation rate. The correction is kept below 0.9 seconds by insertion of the leap-seconds into UTC. Measured and predicted values of the time correction, the leap-seconds and the polar motion are distributed by the International Earth Rotation Service.

The Earth rotates one turn in one *sidereal day* = $2\pi/\psi$ = 23 hours 56 minutes 4 seconds, relative to an inertial frame, and the Sun is seen to move by almost 1 deg/day in the same direction, so the Earth rotates in the mean one turn in 24 hours relative to the Sun direction.

The fact that the duration of one year is not an integer number of days has caused problems for time keeping since the beginning of civilisation. The presently used Gregorian Calendar, introduced by Pope Gregory XIII in the year 1582, defines every fourth year to be a leap-year, except a few integer centuries. The exception does not apply to the year 2000, so one can use the regular four-year cycle from the years 1901 to 2099.

The instantaneous position of the spacecraft can be given in MEGSD as a vector $\bar{r}$ with Cartesian co-ordinates (x, y, z) or spherical co-ordinates (r, s, θ), related by:

$$\bar{r} = \begin{pmatrix} x \\ y \\ z \end{pmatrix} = \begin{pmatrix} r \cos\theta \cos s \\ r \cos\theta \sin s \\ r \sin\theta \end{pmatrix}$$

In general, in astronomy, s and θ are called the right ascension and declination, respectively, of the object. s can also be called the sidereal angle or hour angle of the spacecraft, and r is the absolute value of $\bar{r}$. θ is also the spacecraft subsatellite latitude. The subsatellite longitude λ (Figure C) is the difference between the spacecraft and Greenwich sidereal angles

$$\lambda = s - G$$

We will here count it positive for east and negative for west longitudes. Other sign conventions for λ are, however, used by non-European space agencies. The set (r, λ, θ) constitutes the spherical co-ordinates of the spacecraft in the *Earth-rotating* system. Usually s and λ are adjusted to lie in the interval $(-180°, +180°)$ or $(0°, 360°)$, but there is no generally adopted convention.

It remains to define a co-ordinate system of which the spacecraft is in the centre. A right-handed orthogonal system that will be used here is, Figure D:

- East, forward, tangential along-track in the direction of flight
- North, orthogonal to the orbital plane
- Up, radial out from the Earth

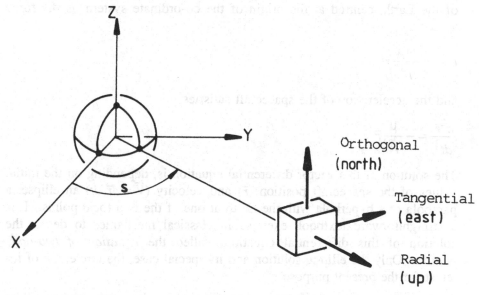

Figure 2.1.D. The spacecraft-oriented co-ordinate system: east, north, up or tangential, orthogonal, radial.

Other definitions of spacecraft oriented systems exist, e.g. east, south, down, which is used by some spacecraft manufacturers. On three-axis stabilised spacecraft the attitude control system often uses the following notation for the control axes:

- Roll axis = along-track
- Pitch axis = orthogonal to orbit
- Yaw axis = radial from Earth

2.2 Elliptical Orbits and Classical Elements

In this section we will further consider the so-called *unperturbed motion*, i.e. an orbit that is only influenced by the gravitational attraction of a spherically symmetric Earth. The co-ordinate system is the previously defined MEGSD. The spacecraft position in this system is represented by the position vector $\bar{r}$, which is a function of time t. The gravitational attraction of the Earth, centred at the origin of the co-ordinate system, is the force vector

$$\bar{F} = -\frac{m\mu}{r^3}\,\bar{r}$$

and the acceleration of the spacecraft satisfies

$$\frac{d^2\bar{r}}{dt^2} = -\frac{\mu}{r^3}\,\bar{r}$$

The solution of this vector differential equation is, depending on the initial values of the spacecraft position ($\bar{r}$) and velocity ($\bar{V} = d\bar{r}/dt$) an ellipse, a parabola or a hyperbola with the Earth at one of the two focal points. It is a straightforward textbook exercise in classical mechanics to derive the solution of this differential equation, called the *equation of two-body motion*. Only the elliptic solution and its special case, the circle, are of interest for the present purpose.

We will only give here some hints on how to obtain the general solution of the differential equation. The first step is to show that the integrated orbit lies in a plane through the centre of the Earth. This is seen from the fact that the angular momentum vector $\bar{r} \times \bar{V}$ is constant since its time derivative becomes zero when the expression for the gravitational acceleration above is inserted for the second derivative of $\bar{r}$.

The differential equation is then expressed in plane polar co-ordinates. The angular component v (= the true anomaly explained below) can easily be integrated by one order. The result is then used to substitute time, as the independent variable, by v in the differential equation for the radial component r, while the latter is replaced by its inverse. The resulting equation is then the well-known harmonic oscillator. Its solution can be written, with the orbital elements a and e (below) as constants of integration:

$$\frac{1}{r} = \frac{1 + e \cos v}{a(1 - e^2)}$$

In general, a vector differential equation of dimension 3 and order 2 needs 6 constants of integration to define a particular solution. A particular solution of the spacecraft equation of motion is called an *orbit* and the the 6 constants of integration can be expressed as 6 *orbital elements* or as a 6-dimensional *state vector*. A state vector consists of the 3 position co-ordinates and the 3 velocity co-ordinates

$$(x, y, z, dx/dt, dy/dt, dz/dt) = (\bar{r}, \bar{V})$$

at a given time t. A time used as reference for the state vector or for orbital elements is called the *epoch*. An important property of an unperturbed orbit is the fact that it lies in a plane through the centre of the Earth and that the ellipse has a constant size, shape and orientation in this plane. Perturbed motion can be described as an ellipse where all these properties, and also the orientation of the plane, are slowly changing with time, but the plane must still pass through the Earth's centre.

The generally used classical orbital elements describe directly the properties of the ellipse. Their mathematical notations, names and units are:

a = semimajor axis (km)
e = eccentricity (dimensionless)
i = inclination (degrees or radians)
Ω = right ascension of the ascending node (degrees or radians)
ω = argument of perigee (degrees or radians)
v = true anomaly (degrees or radians)

All space agencies use this set of elements with the same definition, except that the true anomaly is sometimes replaced by the *mean anomaly*. The mean anomaly, like the *eccentric anomaly*, are mainly used as auxiliary parameters in analytical expression for calculating the time as a function of the true anomaly. They will not be used here since they do not have the same simple geometrical interpretation as the true anomaly.

The point of the orbit where the spacecraft passes closest to the Earth is called *perigee* and the most distant point is called *apogee*. The corresponding distances from the Earth's centre are denoted by r_P and r_A, respectively, Figure A. The definition of the semimajor axis is:

$$a = (r_A + r_P)/2$$

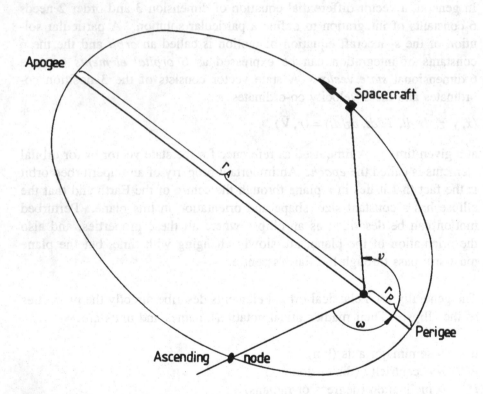

Figure 2.2.A. In-plane orbital elements for an elliptic orbit.

The eccentricity is a dimensionless parameter that expresses how elongated the shape of the ellipse is. It can be obtained by

$$e = r_A/a - 1 = 1 - r_P/a$$

It satisfies the inequality $0 \leq e < 1$, where $e = 0$ means a circular orbit. A near geostationary orbit must have a very small value of e, and for an exactly geostationary orbit it is exactly zero.

The orientation of the plane of the orbit is expressed by the two angles i and Ω. The angle between the orbital plane and the x-y-plane is the inclination i, lying in the interval $0° \leq i \leq 180°$, Figure B. It must be close to $0°$ for a near geostationary orbit and exactly $0°$ for an exactly geostationary orbit.

The ascending node is the point where the orbit crosses the equatorial plane from south to north. The angle between the x-axis and the ascending node is Ω. It is usually defined to lie in the interval $0° \leq \Omega < 360°$. The descending node is the opposite crossing point, from north to south, but it is normally not used as an orbital element.

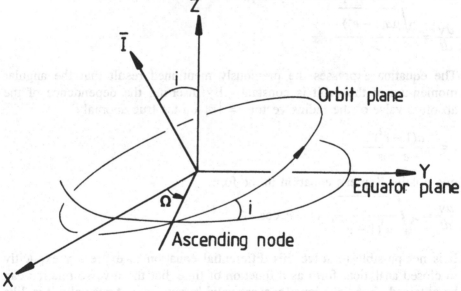

Figure 2.2.B. Orientation of the orbital plane and the out-of-plane orbital elements.

A unit vector orthogonal to the orbital plane, positive with respect to the motion of the spacecraft along its orbit, is parallel to the angular momentum vector and has the components

$$\bar{I} = \begin{pmatrix} \sin i \, \sin \Omega \\ -\sin i \, \cos \Omega \\ \cos i \end{pmatrix}$$

This vector is called the *orbital pole* or the *three-dimensional inclination vector* and its projection on the x-y-plane the *two-dimensional inclination vector*.

The fifth orbital element, ω, gives the orientation, in the orbital plane, of the ellipse with respect to the ascending node, as shown in Figure A. It is the angle from the ascending node crossing to the position of the perigee.

The five orbital elements defined so far are constants for unperturbed orbits, whereas the sixth element, the true anomaly ν, is a function of time. It shows the momentary position of the spacecraft in the ellipse as the angle from perigee, Figure A. The true anomaly satisfies a differential equation that was found empirically by Kepler:

$$\frac{d\nu}{dt} = \frac{\sqrt{\mu a(1 - e^2)}}{r^2}$$

The equation expresses the previously mentioned result that the angular momentum of the orbit is constant. By inserting the dependence of the absolute value of the radius vector $r = |\bar{r}|$ on the true anomaly

$$r = \frac{a(1 - e^2)}{1 + e \, \cos \nu}$$

one obtains Kepler's equation in the form

$$\frac{d\nu}{dt} = \sqrt{\frac{\mu}{a^3(1 - e^2)^3}} \, (1 + e \, \cos \nu)^2$$

It is not possible to solve this differential equation to express ν explicitly in closed analytical form as a function of time, but the inverse relation can be obtained. The equations for ν are valid in radians, but normally it is, like other angles, converted to degrees when a numerical value is given. In one orbital revolution ν increases by 2π radians or 360°. Normally one subtracts multiples of 360° from ν such that it stays between 0° and 360°. The time for a revolution, the *orbital period*, can be calculated to be, independently of the eccentricity:

$$T = 2\pi\sqrt{a^3/\mu}$$

For a geostationary spacecraft the period shall ideally be one sidereal day = $2\pi/\psi$ = 86164 seconds = 23 hours 56 minutes 4 seconds.

It is now possible to express the spacecraft position vector by means of the six orbital elements:

$$\bar{r} = \begin{pmatrix} x \\ y \\ z \end{pmatrix} = \frac{a(1-e^2)}{1+e\cos\nu} \begin{pmatrix} \cos\Omega\,\cos(\omega+\nu) - \sin\Omega\,\sin(\omega+\nu)\,\cos i \\ \sin\Omega\,\cos(\omega+\nu) + \cos\Omega\,\sin(\omega+\nu)\,\cos i \\ \sin(\omega+\nu)\,\sin i \end{pmatrix}$$

One obtains the velocity vector $d\bar{r}/dt$ by taking the derivative of the above equation with respect to t. In this calculation, the derivative of ν with respect to t is obtained from Kepler's differential equation above, whereas the other orbital elements are treated as constants. The set of six equations thus obtained can be inverted, so that the six orbital elements $(a, e, i, \Omega, \omega, \nu)$ are expressed as functions of the spacecraft state vector $(\bar{r}, d\bar{r}/dt)$.

It may seem superfluous to pay so much attention to the definition of orbital elements for exactly elliptical orbits, which do not appear in real life. It is, however, customary to give orbital elements also for perturbed orbits, the so-called *osculating* orbital elements.

The osculating elements at a given instant in time, the *epoch*, are obtained by inserting the true spacecraft position and velocity $(\bar{r}, d\bar{r}/dt)$ obtained from the perturbed motion at the epoch into the formula for the orbital elements of the elliptical motion. In this way, the six orbital elements can be seen as a transformation of the six components of the state vector, with a unique inverse.

In classical celestial mechanics the differential equation of the orbital elements is used for semi-analytical solutions of the perturbed orbit. This transformation of the differential equation has, however, become superfluous since the introduction of digital computers for orbital mechanics. The osculating elements will only be used here as a means of describing an orbit in a more intuitive way than is possible with a state vector.

An exactly geostationary orbit has $i = 0°$, $e = 0$ and $a =$ the geostationary radius. The values of Ω, ω and ν are undefined, but these are only mathematical and not physical singularities. In a computer program where orbital elements are output one can circumvent the problem with a simple IF-statement. However, for orbit determination one must avoid the singularities by, e.g. using the synchronous elements of the next section.

2.3 Linearised Motion and Synchronous Elements

An ideal geostationary spacecraft is at rest with respect to the Earth. For an approximately geostationary orbit it is of interest to find a mathematical expression for the small motion of the spacecraft about the ideal resting position. This is best done by the traditional mathematical method of linearisation (power series expansion to degree one) with respect to the small parameters. One advantage of this linearisation is that one can add linearly the different influences on the orbit: the deviation of the orbital elements from the geostationary values, on the one hand, and the influence of perturbations and manoeuvres, on the other.

This approximate addition of contributions to the spacecraft motion is very useful in helping to visualise, in particular, the effect of manoeuvres. The linear approximation is not needed for direct orbit computation because this can be performed with a high accuracy by direct numerical integration of the differential equation in a computer program.

We start by defining the theoretical position in space for an ideal geostationary spacecraft. Relative to the Earth it is at rest, fixed at zero latitude and with a constant longitude λ_m, where the index m denotes a "mean" longitude. The distance from the centre of the Earth equals the unperturbed geostationary semimajor axis

$$A = \sqrt[3]{\mu/\psi^2} = 42164.2 \text{ km}$$

The corresponding sidereal angle of the ideal position can be expressed as a function of time by:

$$s_m = G + \lambda_m = G_0 + \psi(t - t_0) + \lambda_m$$

In the inertial coordinate system MEGSD, the ideal geostationary position is represented by the rotating vector

$$\bar{r}_m = (A \cos s_m, \ A \sin s_m, \ 0)$$

which has got the constant velocity $V = A\psi$ in the tangential direction defined below. The actual spacecraft motion near the ideal point can be expressed with the help of the three mutually orthogonal unit vectors in the directions: radial, tangential and orthogonal

radial	$= (\cos s_m, \ \sin s_m, \ 0)$
tangential	$= (-\sin s_m, \ \cos s_m, \ 0)$
orthogonal	$= (0,0,1)$

The spacecraft position in the Earth-rotating system is expressed by means of the previously introduced spherical co-ordinates: radius (r), longitude (λ) measured along the tangential direction and latitude (θ) in the orthogonal direction.

The orbit to be studied in this section is still the unperturbed motion. We use the equations from the previous section to linearise with respect to small values of $(e, i, \delta a)$. The deviation of the semimajor axis (a) from the geostationary value (A) is represented by $\delta a = a - A$. The other three elements (Ω, ω, ν) can take any values.

In the previous section the two-dimensional inclination vector was introduced as two components of the three-dimensional inclination vector. Because of the small inclination one can approximate $\sin i \approx i$ in radians and write it as:

$$\bar{i} = (i_x, i_y) = (i \sin \Omega, \ -i \cos \Omega)$$

It has the magnitude $= i$ and points in the direction $\Omega - 90°$, Figure A. Some space centres use another definition of the two-dimensional inclination vector, where it is rotated by $+90°$ to point in the direction of the ascending node Ω. However, our definition has the advantage of a simple geometric interpretation, namely the projection of the orbital pole on the equator plane.

We can also define a two-dimensional eccentricity vector

$$\bar{e} = (e_x, e_y) = [e \cos(\Omega + \omega), \ e \sin(\Omega + \omega)]$$

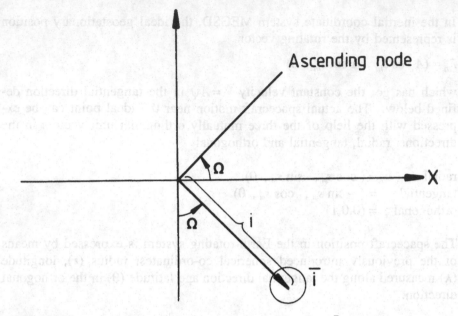

Figure 2.3.A. The two-dimensional inclination vector $\bar{i}$ and its error circle with radius ε_i.

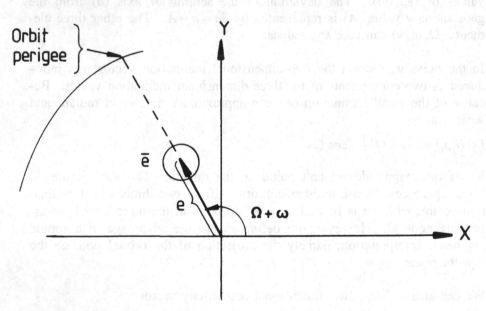

Figure 2.3.B. The eccentricity vector $\bar{e}$ and its error circle with radius ε_e.

which can be visualised as a vector in the x-y-plane with magnitude $= e$ and pointing from the centre of the co-ordinate system in the direction of the orbit perigee, Figure B.

We now put the two second order terms to zero $e^2 \approx 0$ and $e \, \delta a \approx 0$ in the equations of the previous section and obtain the following for r:

$$r = \frac{a(1 - e^2)}{1 + e \, \cos \nu} \approx a(1 - e \, \cos \nu) \approx A + \delta a - A \, e \, \cos \nu$$

In two of the equations for the spacecraft position

$$x/r = \cos \theta \, \cos s = \cos \Omega \, \cos(\omega + \nu) - \sin \Omega \, \sin(\omega + \nu) \, \cos i$$

$$y/r = \cos \theta \, \sin s = \sin \Omega \, \cos(\omega + \nu) + \cos \Omega \, \sin(\omega + \nu) \, \cos i$$

the approximate value $\cos i \approx 1$ is inserted:

$$x/r = \cos \theta \, \cos s \approx \cos(\Omega + \omega + \nu)$$

$$y/r = \cos \theta \, \sin s \approx \sin(\Omega + \omega + \nu)$$

which means that

$$s = \Omega + \omega + \nu$$

In the third component, the linearisations $\sin i \approx i$ and $\sin \theta \approx \theta$ lead to:

$$z/r = \sin \theta = \sin(\omega + \nu) \, \sin i$$

$$\theta \approx i \, \sin(\omega + \nu)$$

One can substitute

$$\nu = s - \Omega - \omega$$

on the right-hand sides of the equations for r and θ to express the spacecraft motion as a function of its sidereal angle s. It is then convenient to insert the eccentricity and inclination vectors as coefficients:

$$r = A + \delta a - A \, e \, \cos(s - \Omega - \omega) = A + \delta a - A \, (e_x \, \cos s + e_y \, \sin s)$$

$$\theta = i \, \sin(s - \Omega) = - i_x \, \cos s - i_y \, \sin s$$

In order to obtain the time-dependence of λ one must solve Kepler's differential equation, from the preceding section, after linearisation for a small eccentricity. We also insert

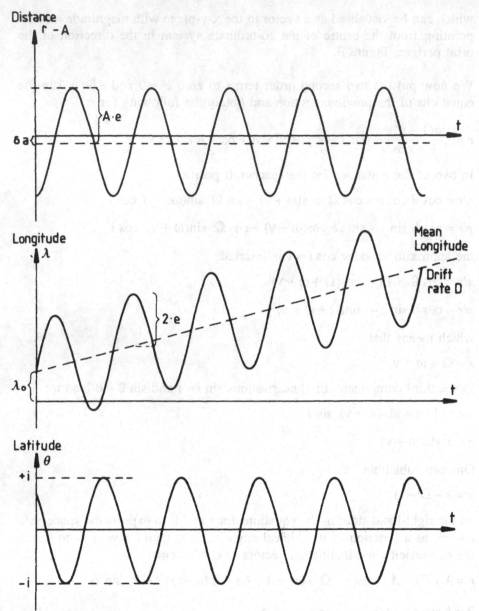

Figure 2.3.C.   Linearised unperturbed spacecraft motion, given as radial distance (r), longitude (λ) and latitude (θ). The sinusoidal libration of each component has the period of one sidereal day. The orbit starts at perigee and covers 5 periods.

$\sqrt{\mu} = \psi A^{3/2}$ to obtain:

$$\frac{d\nu}{dt} \approx \psi(A/a)^{3/2} (1 + e \cos \nu)^2 \approx \psi (1 - 1.5 \, \delta a/A) (1 + 2 \, e \cos \nu) \approx$$

$$\approx \psi (1 - 1.5 \, \delta a/A + 2 \, e \cos \nu)$$

In the following linear solution of the true anomaly as a function of time, the constant of integration t_P is the time of the perigee passage.

$$\nu = \psi (t - t_P) (1 - 1.5 \, \delta a/A) + 2 \, e \sin \psi(t - t_P)$$

The subsatellite longitude is the difference between the sidereal angles of the spacecraft s and the Greenwich meridian G, where G_0 is its value at a suitably selected epoch t_0:

$$\lambda = s - G = \Omega + \omega + \nu - G_0 - \psi(t - t_0) =$$

$$= \Omega + \omega - G_0 + \psi(t_0 - t_P) - 1.5 \, (\delta a/A) \, \psi \, (t - t_P) + 2 \, e \sin \psi(t - t_P)$$

In order to express r and θ as functions of time one can insert the zero-order approximation $\nu \approx \psi(t - t_P)$ into the small terms of the right-hand sides of the previous equations to obtain the first-order time variation:

$$r = A + \delta a - A \, e \cos \nu \approx A + \delta a - A \, e \cos \psi(t - t_P)$$

$$\theta = i \sin(\omega + \nu) \approx i \sin[\omega + \psi(t - t_P)]$$

The time dependence of r, λ and θ is plotted in Figure C. We see that each component contains a term with a sinusoidal *libration* of the same angular frequency, ψ, as the Earth's rotation, i.e. with the period of one sidereal day.

The latitude θ librates symmetrically around zero with an amplitude that is equal to the inclination i. The phase is such that it passes through zero at the ascending and descending nodes since the argument of perigee ω is counted from the ascending node.

The spacecraft distance r from the Earth librates with a phase such that the maxima and minima are located at the apogees and perigees, respectively. The amplitude $= Ae$ and the mean value is off-set by δa from the geostationary value.

The longitude libration has a phase of 90° ahead of r. The amplitude, expressed in radians, is $= 2e$. When converted to kilometres, it is exactly twice the size of the amplitude in the libration of the distance r. This means that an eccentricity of 0.001 causes a longitude libration of $\pm 0.11° = \pm 84$ km in the along-track position. The expression for the longitude is the only component that also contains a term that is linear in time. The coefficient in front of $\psi(t - t_P)$ is called the *mean longitude drift rate*

$$D = - 1.5 \, \delta a/A$$

and is a measure of the deviation between the orbital period and the rotation of the Earth. In the formula above it is expressed as a factor in front of ψ, i.e. D is dimension-less, but for most practical purposes it is converted to degrees per day by multiplication by 361 deg/day. The higher the orbit ($\delta a > 0$), the longer is the orbital period and the more the subsatellite longitude falls behind the Earth's rotation. The result is a mean longitudinal drift westward ($D < 0$), whereas a low orbit ($\delta a < 0$) causes an eastward drift ($D > 0$). The following linear relations can be used for practical operations:

- $\delta a = +1$ km causes the drift-rate $D = -0.0128$ deg/day.

- $D = +1$ deg/day is caused by the off-set $\delta a = -78$ km.

The mean longitude drift rate depends only on the semimajor axis but not on the eccentricity. A non-zero eccentricity causes the longitude to librate as a sine function during the drift, but the mean effect of the libration vanishes when it is taken over an integer number of sidereal days.

Of the expressions for the linear spacecraft motion, the most useful ones are often those where the spacecraft sidereal angle s is used instead of the time as variable on the right-hand sides. Instead of the instantaneous sidereal angle of the spacecraft one can insert

$$s = s_m + \delta s$$

using the previously defined angle s_m. It is clear that δs will always be multiplied with a small factor, so within the approximation of the linearisation it is allowed to put $s = s_m$ everywhere on the right-hand sides. One can then drop the index m in s_m of the future equations and understand s to be the sidereal angle that corresponds to a nominal or mean longitude of the spacecraft. Its value at a selected epoch t_0 becomes:

$$s_0 = \Omega + \omega + \psi(t_0 - t_P)$$

This leads to the linear spacecraft motion in its most useful form:

$$\theta = -i_x \cos s - i_y \sin s$$

$$r = A - A(D/1.5 + e_x \cos s + e_y \sin s)$$

$$\lambda = \lambda_0 + D(s - s_0) + 2e_x \sin s - 2e_y \cos s$$

The first two equations above are the same as the previously derived ones with the spacecraft sidereal angle on the right-hand sides. The last equation has been obtained from the previous expression for λ by the insertion on the right-hand side of

$$\psi(t - t_P) \approx s - \Omega - \omega$$

It contains also a new constant, called the *mean longitude at epoch*, which is defined by:

$$\lambda_0 = \Omega + \omega - G_0 + (1 + D)\psi(t_0 - t_P) = (1 + D)s_0 - G_0 - D(\Omega + \omega)$$

We have used here the mean longitude drift rate D to replace δa since it is often more relevant for describing the properties of a geostationary orbit. One can now use the set of parameters

$$(\lambda_0, D, e_x, e_y, i_x, i_y) = (\lambda_0, D, \bar{e}, \bar{i})$$

as orbital elements instead of the classical elements defined in the previous section. We will call them *synchronous elements*. Various alternative definitions exist in the literature for similar sets of elements, sometimes also known as equinoctial elements. There is no generally used set of standard elements in this case, in contrast to the situation with the classical elements.

The synchronous orbital elements are defined as osculating elements for perturbed orbits in a manner analogous to that employed for the classical elements. For some purposes one is more interested in *mean elements*. There are various ways of defining mean elements, but here we will define them as the arithmetic mean of the osculating synchronous elements, averaged over one sidereal day. It is important to take the average over a whole, but not a fraction of, a sidereal day. Otherwise the result would be misleading because of the short-term perturbations on the orbit.

The synchronous elements $\bar{i}$ and $\bar{e}$ can be considered as Cartesian components of the polar representations $(i, \Omega - 90°)$ and $(e, \Omega + \omega)$ respectively. It is not meaningful to take the mean values of Ω and ω when i and e are close to zero, so we will not calculate mean values of the classical elements for geostationary orbits. Still, one can always obtain mean classical elements by converting the mean synchronous elements through the above transformations in the reverse direction.

Such mean elements are often useful on scheduling prints for the spacecraft Operator as shown in Section 5.2. Synchronous elements are used mostly in analytical formulae of orbital motion but less often for input and output to computer programs. In the former case, the angles are expressed in radians and in the latter case in degrees.

To express variations in the spacecraft position, also in latitude and longitude, in kilometres relative to its mean longitude one needs to multiply the two angles, in radians, by A to obtain:

$$A\theta = -A(i_x \cos s + i_y \sin s)$$

$$r - A = -A(D/1.5 + e_x \cos s + e_y \sin s)$$

$$A(\lambda - \lambda_m) = A(\lambda_0 - \lambda_m) + AD(s - s_0) + 2A(e_x \sin s - e_y \cos s)$$

Figure D shows an orbital motion projected on the equator plane $[r - A, A(\lambda - \lambda_m)]$. The orbit is the same as in Figure C, with a slow east drift. The eccentricity always causes a negative rotation in the equator plane with the motion along an ellipse that is $4Ae$ long, in the tangential direction, and $2Ae$ wide, in the radial direction.

Figure E shows the projection on the meridian plane $[r - A, A\theta]$ of orbital motion caused by eccentricity and inclination. The relative phase between the sinusoidal librations is the argument of perigee ω. When $\omega = 90°$ or $270°$ the inclination and eccentricity vectors are antiparallel or parallel, respectively, and the motion is along a straight line. Between these two values the motion is elliptic, rotating in the positive sense as shown. The other values of ω cause motions along the same ellipses but in the negative sense.

The effect of the eccentricity and inclination on the orbital motion in the local horizontal plane $[A(\lambda - \lambda_m), A\theta]$ is the same as the curves in Figure E, but with ω increased by 90°. The rotation is in the positive sense when $180° < \omega < 360°$ and negative when $0° < \omega < 180°$. Figure F shows an example of the latter motion combined with an east longitude drift.

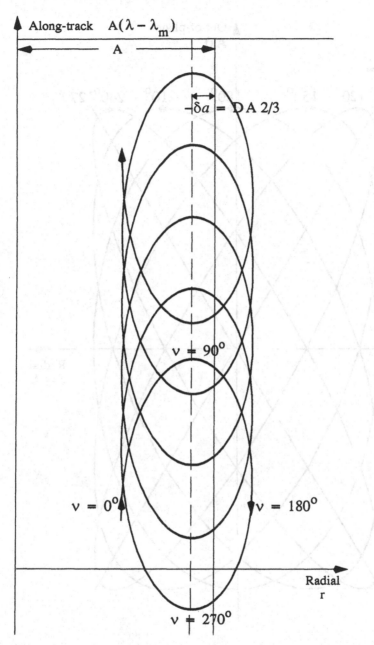

Figure 2.3.D. The same orbital motion as in figure C projected on the equator plane. The true anomaly ν is marked for the first orbital revolution.

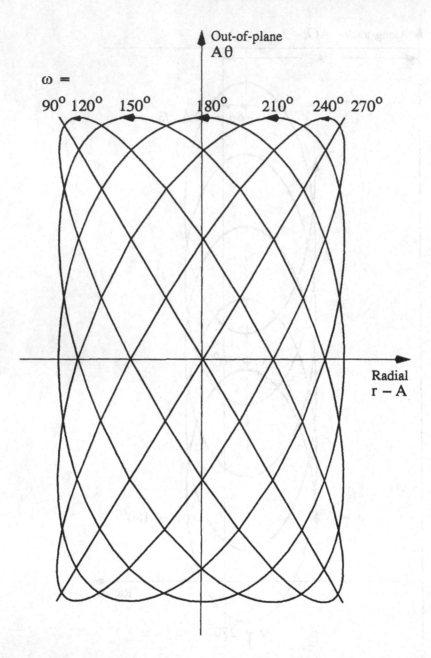

Figure 2.3.E. Orbital motion projected on the meridian plane for different values of the argument of perigee ω as indicated.

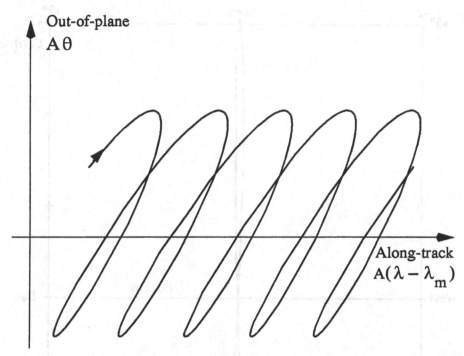

Figure 2.3.F. Orbital motion with an east drift during 5 sidereal days projected on the local horizontal plane. The argument of perigee $\omega = 20°$.

In many popular publications describing the geostationary orbit the groundtrack, i.e. the orbital motion in the local horizontal plane, is said to perform a "figure of eight", aligned in the north-south direction. This result is obtained if one continues the power series expansion for λ with a second-order term in i in radians:

$$0.25 \, i^2 \, \sin 2[\psi(t - t_P) + \omega]$$

The effect is very small, as seen in Figure G. The longitude term with half the orbital period has the coefficient, expressed in degrees, = 0.0044°, 0.017°, 0.039°, respectively, when $i = 1°$, 2°, 3°. The effect on the longitude is noticeable only when the longitude drift, during one day, and the eccentricity librations lie far below. Such is seldom the case, however, since it is unusual to find a mission where a relatively high inclination is combined with a requirement to keep the eccentricity and longitude drift rate very low.

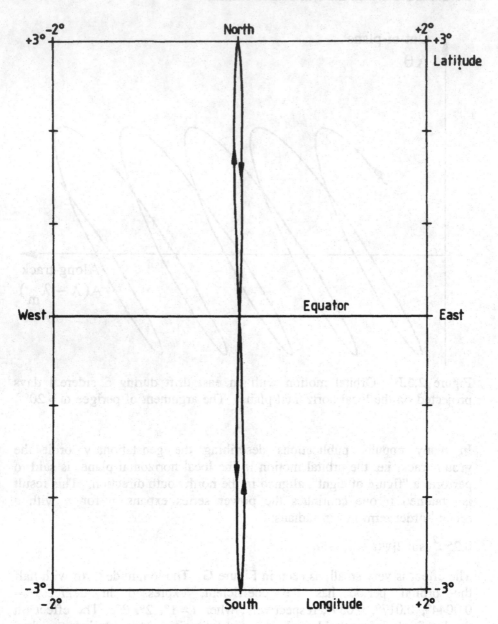

Figure 2.3.G. "Figure-of-eight" subsatellite track. The longitude drift = 0, $e = 0$, $i = 3°$.

The error ε_D of the mean longitude drift rate D also stays approximately constant in time, whereas the error of the mean longitude increases linearly in time by the factor ε_D. Figure H shows the envelopes of the error regions for the mean longitude, the instantaneous longitude and the latitude. The half-width of the maximum error region of the instantaneous longitude is obtained by

$$\varepsilon_\lambda = \varepsilon_{\lambda 0} + \varepsilon_D \psi(t - t_0) + 2\varepsilon_e$$

The next task is now to express, in the linear approximation, the spacecraft velocity in km/s by the time derivatives of the position. For this one needs the time derivative of the nominal sidereal angle

$$\frac{ds}{dt} = \psi$$

The following expressions give the radial, tangential and orthogonal components of the spacecraft velocity relative the rotating Earth.

$$V_r = \frac{dr}{dt} = V(e_x \sin s - e_y \cos s)$$

$$V_t = A\frac{d\lambda}{dt} = V(D + 2e_x \cos s + 2e_y \sin s)$$

$$V_o = A\frac{d\theta}{dt} = V(i_x \sin s - i_y \cos s)$$

The velocity in the inertial frame is obtained by adding the velocity of the ideal geostationary position $V = A\psi$ to the tangential component. The transformation to MEGSD can be done by rotation around the z-axis by the angle $-s$.

The accuracy with which the orbit is known, from the orbit determination, is best expressed by giving the estimated error, ε, for the synchronous elements. The error intervals for the eccentricity and inclination vectors can be expressed as circles with radii ε_e and ε_i, respectively, around the measured elements $\bar{e}_0$ and $\bar{i}_0$, Figures A and B.

$$|\bar{e} - \bar{e}_0| \le \varepsilon_e \quad ; \quad |\bar{i} - \bar{i}_0| \le \varepsilon_i$$

One can use ε_e and ε_i, with the same definitions, also to express the errors of the eccentricity and inclination in connection with the classical elements. This eliminates the need for explicit error intervals for Ω and ω.

It will be shown later that the natural perturbations induce a drift of $\bar{e}$ and $\bar{i}$ in the x-y-plane. According to the linear approximation, used in this section, there is no cross-coupling between the errors of the orbital elements and the natural perturbations. The time evolution of the error regions for $\bar{e}$ and $\bar{i}$ is then very simple: the size and shape of the error circles stay constant, but the circles move in the x-y-plane according to the same law as the drift of $\bar{e}$ and $\bar{i}$.

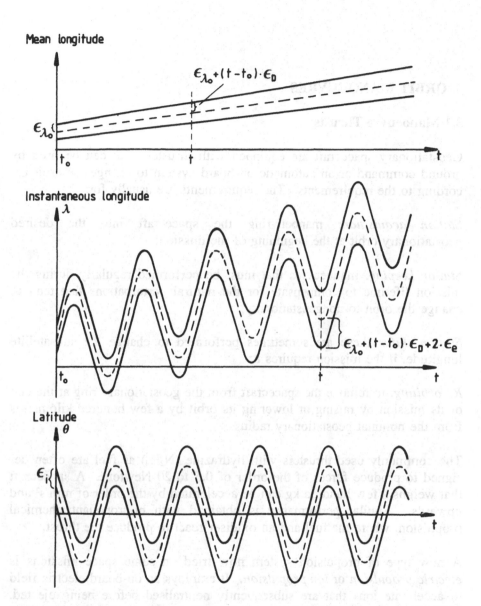

Figure 2.3.H. Propagation of errors ($\varepsilon_{\lambda 0}$, ε_D, ε_e, ε_i) resulting from the determination of the orbital elements (λ_0, D, $\overline{e}$, $\overline{i}$). The error regions of the mean and instantaneous longitudes increase linearly with time, and the error in latitude remains constant.

3. ORBIT MANOEUVRES

3.1 Manoeuvre Thrusts

Geostationary spacecraft are equipped with thrusters that can be fired by ground command or an automatic on-board system to change the orbit according to the requirements. The requirements are usually for:

Station acquisition: manoeuvring the spacecraft into the desired geostationary orbit at the beginning of the mission;

Station keeping: manoeuvres that must be performed regularly during the mission lifetime to compensate for the natural perturbations that tend to change the orbit to nongeostationary;

Station shifts: these are sometimes performed to change the subsatellite longitude, if the mission requires it;

Re-orbiting: to remove the spacecraft from the geostationary ring at the end of its mission by raising or lowering its orbit by a few hundred kilometres from the nominal geostationary radius.

The commonly used thrusters with hydrazine (N_2H_4) as fuel are often designed to produce forces of the order of 0.5 to 20 Newtons. A spacecraft that weighs a few hundred kg can be accelerated by the order of mm/s^2 and upwards. Similar performance is obtained with bi-propellant chemical propulsion, where the fuel and an oxidiser react to produce the thrust.

A new type of propulsion system now tried on some space missions is *electric propulsion* or *ion propulsion*. It employs an on-board electric field to accelerate ions that are subsequently neutralised before being ejected. The advantage is that more ΔV relative to the launch mass can be obtained for missions of long duration because only a small amount of mass is ejected with a high velocity, typically 30 km/s. This is combined with the fact that the accelerating power is obtained from the solar cells instead of being carried on-board as chemicals. The disadvantage is the very low force, of the order of 0.01 Newton, leading to thrust times that are a hundred times longer than for chemical propulsion for producing the same effect on the orbit.

Thruster burns used for geostationary orbit control are either orthogonal to the orbital plane or tangential to the orbit, Figure A; burns in radial direction are hardly ever used. However, for practical design reasons it may sometimes be difficult to mount the thrusters on the spacecraft to produce the desired thrust direction. The force vector has to pass through the mass centre of the spacecraft in order not to produce any torque. A torque can also be avoided by the firing of two thrusters simultaneously. A further requirement is that antennas, solar panels or other protruding equipment must not be damaged by the exhaust plumes. The result may be a compromise where the force has a component in an undesired direction. This decreases the efficiency of the manoeuvre and may also have to be compensated for by additional manoeuvres.

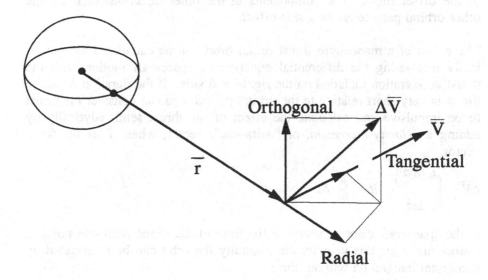

Figure 3.1.A. A velocity increment $\Delta \overline{V}$ from a spacecraft manoeuvre thrust, split into the three orthogonal components: radial, tangential and orthogonal to the orbital plane.

A manoeuvre that produces a force orthogonal to the orbital plane is called
an out-of-plane thrust, a north-south thrust or an inclination manoeuvre. It
is used to change the orientation of the plane of the orbit, i.e. it changes the
inclination and the ascending node. A thrust tangential to the orbit is called
an along-track thrust, an east-west thrust or a longitude manoeuvre. It
changes the orbital semimajor axis, the longitude drift rate and the eccentricity vector. A radial thrust changes only the eccentricity vector, but only
half as much as a tangential thrust of the same size, so it is not cost effective
to manoeuvre the orbit eccentricity in this way. The radial and the along-
track thrusts can both be referred to as in-plane manoeuvres.

The effect of a force that is not aligned to any of these three orthogonal
directions can be calculated by splitting the vector into the three components
radial, tangential and orthogonal and superimpose the effect from each
separate component on the spacecraft motion. If the spacecraft orientation
or the thruster mounting is such that the desired direction cannot be obtained
directly, the efficiency of the burn is reduced by a factor equal to the cosine
of the offset angle. The components in the other directions then change
other orbital parameters as a side-effect.

The effect of a manoeuvre thrust on an orbit can be calculated by numerically integrating the differential equation of spacecraft motion with the
thrust acceleration included on the right-hand side. If the duration Δt of the
thrust is very short relative to the orbital period, one can consider the thrust
to be impulsive and calculate the effect of the thrust semianalytically by
adding a *velocity increment*, or "delta-vee", vector, where $\overline{F}$ is the force
vector,

$$\Delta \overline{V} = \int_{t_b - \Delta t/2}^{t_b + \Delta t/2} \frac{\overline{F}}{m}\, dt \approx \frac{\overline{F}}{m}\, \Delta t$$

to the spacecraft velocity vector at the time of the thrust burn mid-point t_b.
Before and after this velocity discontinuity the orbit can be propagated by
numerical integration without thrust.

If the orbit state vector just before the thrust mid-point was $(\overline{r}, \overline{V})$ the state
vector just after the thrust will be $(\overline{r}, \overline{V} + \Delta \overline{V})$. By inserting these state
vectors into the formula for the orbital elements one can calculate the effects
of $\Delta \overline{V}$ on the elements. This method of introducing a discontinuity into the
velocity part of the state vector can often be used for thrusts lasting up to
an hour as long as the $\Delta \overline{V}$ is allocated to the mid-point time of the thrust.

In order to perform the conversion to and from the orbital elements one shall in principle use the exact definition in Section 2.2, which requires a computer program for the numerical calculations. On the other hand, it is easy to give an approximate, linearised expression for the change in the synchronous elements of Section 2.3 from the thrust $\Delta \overline{V}$. We express the three components of $\Delta \overline{V}$ in the radial, tangential and orthogonal directions by, respectively, $(\Delta V_r, \Delta V_t, \Delta V_o)$. The change of the synchronous elements from the $\Delta \overline{V}$ is expressed by a "Δ" in front of the element, so that the elements just before and just after the thrust are, respectively,

$$(\lambda_0, D, e_x, e_y, i_x, i_y)$$

$$(\lambda_0 + \Delta\lambda_0, D + \Delta D, e_x + \Delta e_x, e_y + \Delta e_y, i_x + \Delta i_x, i_y + \Delta i_y)$$

The fact that the instantaneous spacecraft position does not change at the moment of the impulsive thrust imposes the following condition on the change in the synchronous elements, according to the equations of linearised spacecraft motion in Section 2.3. Here we denote by s_b the spacecraft sidereal angle at the thrust time t_b.

$$0 = \Delta r = -A(2/3 \, \Delta D + \Delta e_x \cos s_b + \Delta e_y \sin s_b)$$

$$0 = \Delta\lambda = \Delta\lambda_0 + \Delta D(s_b - s_0) + 2\Delta e_x \sin s_b - 2\Delta e_y \cos s_b$$

$$0 = \Delta\theta = -\Delta i_x \cos s_b - \Delta i_y \sin s_b$$

We next insert the change in the spacecraft velocity, using the linearised velocity equations in Section 2.3:

$$\Delta V_r = V(\Delta e_x \sin s_b - \Delta e_y \cos s_b)$$

$$\Delta V_t = V(\Delta D + 2\Delta e_x \cos s_b + 2\Delta e_y \sin s_b)$$

$$\Delta V_o = V(\Delta i_x \sin s_b - \Delta i_y \cos s_b)$$

These equations will be solved in Sections 3.2 and 3.3 to show the approximate effect of manoeuvre thrusts on the synchronous elements. This linearisation is justified for approximate calculations since typical ΔVs used for station keeping manoeuvres are of the order of 0.1 to 50 m/s, which is small with respect to the orbital velocity of 3 km/s.

In operational practice the spacecraft operator calculates by means of a computer program what $\Delta \overline{V}$s (time, size and direction) are needed to change the latest determined orbit to whatever new orbit is desired. After this, a model of the on-board thruster system is applied to calculate which thrusters shall be switched on for how long in order to obtain the desired $\Delta \overline{V}$. Such a model can often be very simple, consisting of only a constant acceleration (in m/s^2) with which to divide the ΔV (in m/s) in order to obtain the number of seconds to fire the thruster. The first time a thruster is fired in space one must use a value that is based upon the manufacturer's specifications and the estimated spacecraft mass, although this usually does not produce a very accurate prediction of the performance.

Thrusters, and particularly hydrazine thrusters, are highly individualistic and their performance in space often differs from pre-flight measurements because of the difficulties to reproduce the space environment in a ground laboratory. The first time a hydrazine thruster is used in space the error can be as high as 10% to 15% and in one case (ECS-2 launched in 1984) was found to be almost 50% low. Thrusters for bi-liquid propellants or cold gas are more accurate in the predictability of the thrust size. A further source of error in the $\Delta \overline{V}$ is that, in order to keep their balance during an orbit manoeuvre, three-axis stabilised spacecraft often must modify (off-modulate) some of the thrusting or fire additional attitude control thrusters by an automatic on-board program. This changes the total $\Delta \overline{V}$ by an amount which is often not telemetered to ground.

The following simple mode of in-flight thruster calibration can be used for improving the predictability of the ΔV for subsequent firings. When enough tracking data has been collected one determines the new orbit to find out what the actual ΔV was. By dividing it (m/s) with the time the thruster was fired (seconds) one obtains the new acceleration (m/s^2) that can be used to model the next firing of the same thruster. If this is carried out after each major manoeuvre it will automatically account for the combined effects of the gradually decreasing spacecraft mass and thruster efficiency during the whole mission.

Usually, the fuel mass used for one manoeuvre burn is small, so one can consider the spacecraft mass to be constant during the thrusting. The exponential rocket equation should be used only for the accumulated effect of many thrusts. It is useful to express the spacecraft fuel load in m/s, instead of kg, to show the sum of all velocity increments that can be given to the spacecraft during its mission lifetime. The fuel budget can be specified to contain a certain amount of m/s for station acquisition, others for in-plane and out-of-plane station keeping, etc.

One aspect of manoeuvring is how accurate the timing of the burn execution shall be. Since the spacecraft moves an angle of $\delta s = \psi \delta t$ in its orbit during a delay δt of the start time, the effect of a planned manoeuvre can be described as a reduction by the factor $\cos \delta s$. A delay of half an hour only reduces the effect of a manoeuvre by 1% on the desired orbital element.

3.2 Inclination Manoeuvres

An inclination or north-south thrust is used to change the plane of the orbit, which means that the inclination vector and the classical elements (i, Ω) are changed. A thrust is called a north thrust if it is executed by a thruster on the south surface of the spacecraft (Figure A), giving a $\Delta \overline{V}$ directed northwards. The opposite type of thrust is called a south thrust.

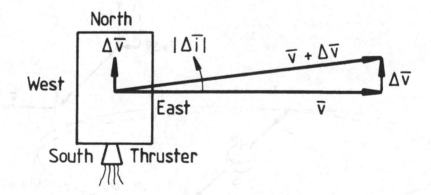

Figure 3.2.A. Execution of a north thrust with a thruster on the south surface of the spacecraft. The velocity increment vector $\Delta \overline{V}$ is added to the spacecraft's orbital velocity $\overline{V}$.

The angular momentum of the orbit is, per unit mass of the spacecraft, the vector $\bar{r} \times \bar{V}$, which has the magnitude of about $13 \times 10^4 \, km^2/s$ and is parallel to the three-dimensional inclination vector $\bar{I}$. A small $\Delta\bar{V}$ parallel to $\bar{I}$ given to the spacecraft at the position $\bar{r}$ changes the angular momentum by $\bar{r} \times \Delta\bar{V}$. The angle between the old and the new orbital planes is approximately (in radians) $\Delta V/V$, Figures A and B. The two orbital planes intersect at the line through $\bar{r}$ and $-\bar{r}$. A thrust of $-\Delta\bar{V}$ at position $-\bar{r}$ has the same effect as the thrust $\Delta\bar{V}$ at $\bar{r}$, so by selecting the time of the thrust either a north or a south thrust can be used for the same result.

Ideally, an inclination manoeuvre $\Delta\bar{V}$ should be orthogonal to the plane bisecting the planes of the old and the new orbits. If this is not the case, one will obtain an in-plane component of $\Delta\bar{V}$ that may interfere with the longitude manoeuvres, as will be described later.

We use the linearised equations for a small orthogonal ΔV and i from Section 3.1. In this section we remove the index and write $\Delta V = \Delta V_o$. We now solve $(\Delta i_x, \Delta i_y)$ from the following two equations

$$0 = -\Delta i_x \cos s_b - \Delta i_y \sin s_b$$

$$\Delta V = V(\Delta i_x \sin s_b - \Delta i_y \cos s_b)$$

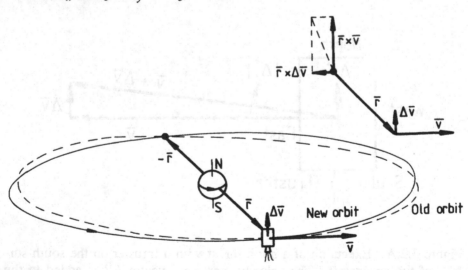

Figure 3.2.B. Change of the orbital plane by a north thrust, $\Delta V > 0$. The dashed line indicates the orbit before the thrust.

to obtain the change of the two-dimensional inclination vector, Figure C:

$$\Delta \bar{i} = \bar{i}_{new} - \bar{i}_{old} = \frac{\Delta V}{V} \begin{pmatrix} \sin s_b \\ -\cos s_b \end{pmatrix}$$

The time of day of the thrust is so chosen such that the spacecraft sidereal angle s_b, at the time of the thrust, gives the desired direction of $\Delta \bar{i}$. ΔV is defined to be positive for a north thrust and negative for a south thrust. By inserting this expression into the linearised orbital motion of Section 2.3 we now obtain the linearised evolution in the spacecraft latitude after the manoeuvre:

$$\Delta \theta = (\Delta V/V) \sin(s - s_b) \quad \text{when } s > s_b$$

This is the difference between the latitude resulting from the manoeuvre and what the latitude would have been without the manoeuvre. Before the manoeuvre, $\Delta \theta$ is of course $= 0$. The absolute value of the inclination change $|\Delta \bar{i}|$ is the same as the angle between the old and the new orbital planes. One can obtain an angle of 1° from a thrust of 53.7 m/s. Conversely, a thrust of 1 m/s only gives the angle 0.0186°, corresponding to an amplitude of the out-of-plane deviation of $= \Delta V/\psi = 13.7$ km.

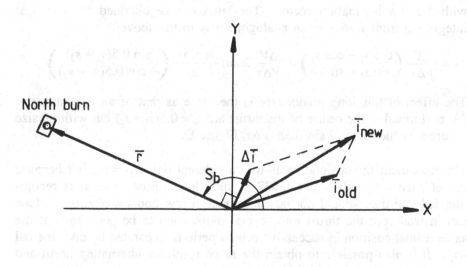

Figure 3.2.C. Linearised representation of the change of the orbital plane by use of the two-dimensional inclination vector.

The new inclination depends upon the the start inclination and the right ascension of the ascending node and on the sidereal angle of the thrust relative to the latter:

$$i_{new} = \sqrt{i_{old}^2 + (\Delta V/V)^2 + 2(\Delta V/V)i_{old}\,\cos(s_b - \Omega_{old})}$$

If the thrusters are weak it may be necessary to perform a thrust that is so long (Δt) that the spacecraft sidereal angle changes an appreciable amount $\Delta s = \psi \Delta t$ during the thrust even though the orbital change due to the thrust is small. We denote the sidereal angle at the thrust start by s_1 and at the end by $s_2 = s_1 + \Delta s$.

In order to obtain the evolution of the spacecraft latitude during the thrusting one can perform an analytical integration of the above formula as follows. Instead of a single ΔV at one fixed angle s_b we have an infinite number of thrusts $(\Delta V/\Delta s)ds_b$ at a continuum of different s_b. The effect at time t, corresponding to the angle s, is the sum, or rather integral, from s_1 to s:

$$\Delta\theta = \frac{\Delta V}{V\Delta s}\int_{s_1}^{s}\sin(s - s_b)\,ds_b = \frac{\Delta V}{V\Delta s}[1 - \cos(s - s_1)] \quad \text{when } s_1 \le s \le s_2$$

As soon as the thrust stops the spacecraft follows a free flight again but with the new inclination vector. The latter can be obtained by analytical integration from s_1 to s_2 in an analogous way to the above:

$$\Delta\bar{i} = \frac{\Delta V}{V\Delta s}\begin{pmatrix}\cos s_1 - \cos s_2 \\ \sin s_1 - \sin s_2\end{pmatrix} = \frac{\Delta V}{V\Delta s}\,2\sin\frac{s_2 - s_1}{2}\begin{pmatrix}\sin 0.5(s_1 + s_2) \\ -\cos 0.5(s_1 + s_2)\end{pmatrix}$$

The effect of this long manoeuvre is the same as that of an instantaneous ΔV performed at the centre of the thrust arc $s_b = 0.5(s_1 + s_2)$ but with its size reduced by the factor $2\sin(0.5\Delta s)/\Delta s$, Figure D.

The maximum thrust duration is half a sidereal day ($\Delta s = \pi$), but because the efficiency is only $= 2/\pi = 0.637$ at such long thrust times it is recommended that they should not last more than a few hours at the most. One can instead split the thrust into several parts, each to be performed at the same orbital position in successive orbital periods, separated by one sidereal day. It is also possible to obtain the same result by alternating north and south thrusts twice per sidereal day.

For inclination manoeuvres there are no multi-thrust sequences in the same way as for longitude manoeuvres. The only reason to perform many thrusts in a sequence is when the thrusters are so weak that one firing would not be enough. The resulting effect the orbit is obtained in the linearised approximation by vector addition of each $\Delta\bar{i}$ to the original vector $\bar{i}$.

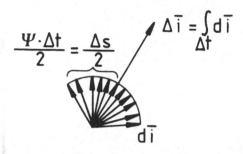

Figure 3.2.D. Efficiency loss of a long inclination thrust during which the spacecraft sidereal angle moves by Δs.

3.3 Single In-plane Manoeuvre

A longitude or east-west thrust changes both the longitude drift rate and the eccentricity of the orbit. A three-axis stabilised spacecraft that has one side always turned towards the Earth can obtain a push in the direction of flight by a thruster placed on the westward surface, Figure A. This is called an east thrust and the corresponding ΔV is counted positive. The opposite case is called a west thrust, with a negative ΔV. On spin-axis stabilised spacecraft, where the spin axis is orthogonal to the orbital plane (Figure B), an east or west thrust is obtained by firing pulses during a part of the spin revolution. The accumulated effect of the pulses on the spacecraft can be represented by a ΔV if the total manoeuvre duration is short with respect to an orbital revolution.

A single tangential thrust that can be considered as an instantaneous impulse changes instantaneously the flight velocity of the spacecraft but not its position. Figure C shows an initially circular orbit that is exposed to an east thrust. The ΔV raises the orbit everywhere except at the place where the burn was made. The new orbit will keep the original height at this point and the new perigee will be located there. The semimajor axis and the eccentricity of the new elliptic orbit are modified such that this side of the orbit remains fixed and the other side is raised by an amount proportional to the ΔV. A west thrust, with a negative ΔV, would lower the orbit in an analogous way to put the new apogee at the point of the thrust.

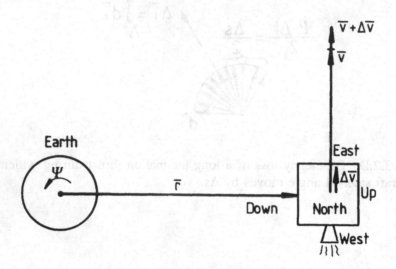

Figure 3.3.A. An east thrust with a three-axis stabilised spacecraft, seen from the north.

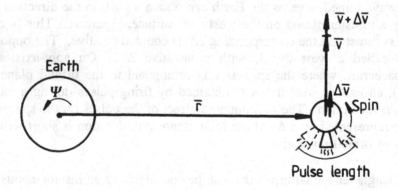

Figure 3.3.B. An east thrust in pulsed mode with a spin-stabilised spacecraft, seen from the north.

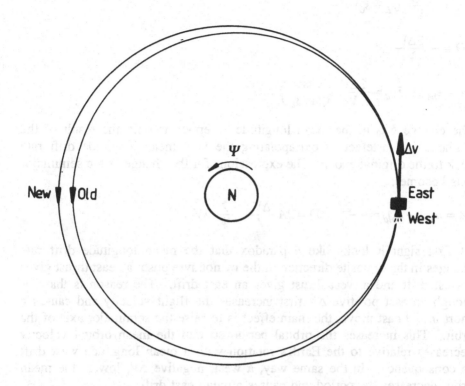

Figure 3.3.C. Change in orbit due to an east thrust.

We first consider an instantaneous tangential $\Delta V = \Delta V_t$ at at time t_b, corresponding to a spacecraft sidereal angle s_b. According to the linearised expressions in Section 3.1 the change of the synchronous elements must satisfy the four equations:

$$0 = -A(2/3 \, \Delta D + \Delta e_x \cos s_b + \Delta e_y \sin s_b)$$

$$0 = \Delta\lambda_0 + \Delta D(s_b - s_0) + 2\Delta e_x \sin s_b - 2\Delta e_y \cos s_b$$

$$0 = V(\Delta e_x \sin s_b - \Delta e_y \cos s_b)$$

$$\Delta V = V(\Delta D + 2\Delta e_x \cos s_b + 2\Delta e_y \sin s_b)$$

The solution for the four elements ($\Delta\lambda_0$, ΔD, Δe_x, Δe_y) then becomes:

$$\Delta\lambda_0 = \frac{3\Delta V}{V}\,(s_b - s_0)$$

$$\Delta D = -\frac{3\Delta V}{V}$$

$$\Delta\bar{e} = \bar{e}_{new} - \bar{e}_{old} = \frac{2\Delta V}{V}\begin{pmatrix}\cos s_b\\ \sin s_b\end{pmatrix}$$

The change $\Delta\lambda_0$ in the mean longitude of epoch is only the result of the mathematical artefact of extrapolating the new mean longitude drift rate back to the original epoch. The expressions for the change of the semimajor axis becomes:

$$\Delta a = a_{new} - a_{old} = -\frac{2A}{3}\,\Delta D = 2A\,\frac{\Delta V}{V} = \frac{2}{\psi}\,\Delta V$$

At first sight it looks like a paradox that the mean longitude drift rate changes in the opposite direction of the manoeuvre push: an east thrust gives a west drift and a west thrust gives an east drift. The reason is that, although an east positive ΔV first increases the flight velocity and causes a short initial east move, the main effect is to raise the semimajor axis of the orbit. This increases the orbital period so that the mean orbital velocity decreases relative to the Earth's rotation with a mean longitude west drift as consequence. In the same way, a west, negative ΔV, lowers the mean orbit, decreases the period and causes a mean east drift.

The following linear relations are useful for practical operations:

- $\Delta V = +1$ m/s causes a $\Delta a = +27.4$ km, $\Delta D = -0.352$ deg/day and $|\Delta\bar{e}| = 0.000650$

- $\Delta D = +1$ deg/day is obtained by a thrust $\Delta V = -2.84$ m/s, which causes a $\Delta a = -78$ km and $|\Delta\bar{e}| = 0.00185$

The change in the eccentricity vector $\Delta\bar{e}$ is parallel to the projection of the spacecraft position vector on the x-y-plane at the moment of the thrust, Figure H. The change in orbital height at the point opposite to the thrust equals twice the change in semimajor axis.

Following are the linearised expressions, as functions of the sidereal angle s, for the changes in r and λ caused by a thrust at the sidereal angle s_b. The equations are valid also when the initial orbit was not perfectly geostationary, and in the presence of natural orbital perturbations, if Δ is taken to mean the differences between the parameters after the thrust compared to what the values would have been at the same times without the thrust.

$$\Delta r = \frac{2\Delta V}{\psi}\left[1 - \cos(s - s_b)\right] \quad \text{when } s > s_b$$

$$\Delta \lambda = \frac{\Delta V}{V}\left[4\sin(s - s_b) - 3(s - s_b)\right] \quad \text{when } s > s_b$$

The longitude λ does not change instantaneously from the thrust but changes with time as a result of the new mean longitude drift rate and eccentricity. Figure D shows the longitude as a function of time for the orbit of Figure C exposed to an east thrust with $\Delta V = +1$ m/s. The longitude of the initially perfectly geostationary orbit is constant $= \lambda_0$ up to the moment of the thrust. At the instant of the thrust, λ gets a push towards east and, after 2 hours 45 minutes reaches a maximum of $\lambda_0 + 0.0089°$, corresponding to an along-track deviation of $+6.55$ km. It then reverses to pass λ_0 at 4 hours 51 minutes after the thrust and continues with a mean west drift but with superimposed sinusoidal librations that cause a small east drift during 5 hours 30 minutes each sidereal day.

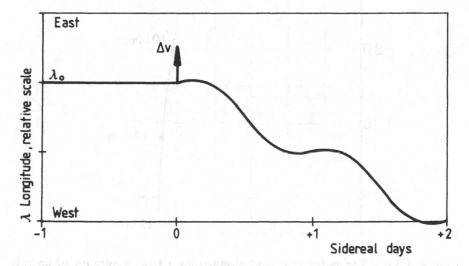

Figure 3.3.D. Change in longitude drift due to an east thrust. The vertical scale division is proportional to the thrust: $0.352°$ for $\Delta V = 1$ m/s.

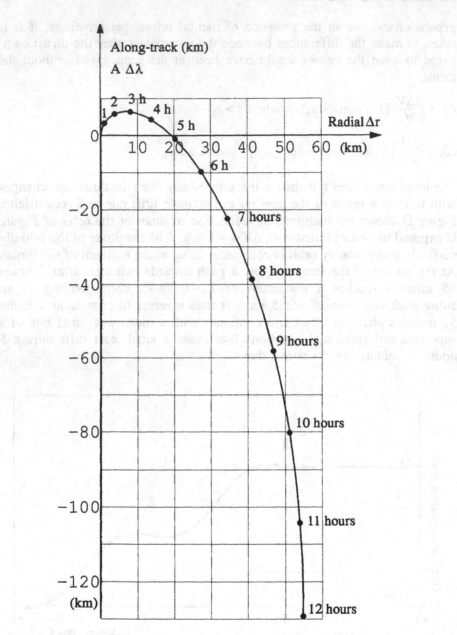

Figure 3.3.E. The radial (Δr) and along-track ($A\Delta\lambda$) evolution of an orbit during the first 12 hours after a 1 m/s east thrust.

Figure E shows the first 12 hours of the same orbital evolution plotted as Δr versus $A\Delta\lambda$ in km. It should be noted that the displacement along-track is very much larger than in the radial direction. The former increases by -130 km every 12 hours, whereas the latter librates with the maximum excursion of +54.8 km after 12 hours. This high sensitivity of the along-track position for a given ΔV compared to the radial, and also the out-of-plane, spacecraft position is of great importance for the effect of thrust errors, as will be seen later.

The consequence of using long manoeuvre thrusts that cannot be directly approximated by a single ΔV is analogous to the situation for inclination thrusts. However, depending on the spacecraft design the ΔV may remain tangential to the orbit during the whole thrust or only at one instant. The former situation applies to three-axis stabilised spacecraft that constantly turn one side towards the Earth. In this case the acceleration can remain tangential to the orbit during the whole thrust, Figure F.

We assume that the long tangential thrust of duration $\Delta t = \Delta s / \psi$ starts at the sidereal angle s_1 and ends at $s_2 = s_1 + \Delta s$. One can obtain the expression for the two in-plane spacecraft position components during the thrusting by analytical integration of the equations for the impulsive thrust with respect to s_b from s_1 to s in the same way as in Section 3.2.

We have, when $s_1 \leq s \leq s_2$

$$\Delta r = \frac{2\Delta V}{\psi \Delta s} \int_{s_1}^{s} [1 - \cos(s - s_b)] \, ds_b = \frac{2\Delta V}{\psi \Delta s} [s - s_1 - \sin(s - s_1)]$$

$$\Delta \lambda = \frac{\Delta V}{V \Delta s} \int_{s_1}^{s} [4 \sin(s - s_b) - 3(s - s_b)] \, ds_b =$$

$$= \frac{\Delta V}{V \Delta s} [4 - 4 \cos(s - s_1) - 1.5 \, (s - s_1)^2]$$

After the end of the thrust the spacecraft continues in a free drift orbit with the new orbital elements. The change in semimajor axis and mean longitude drift rate depend only on the total ΔV, regardless of the thrust length, as long as the thrust everywhere is tangential to the orbit. The effect on the eccentricity vector is the same as of an impulsive ΔV reduced by the factor $2 \sin(0.5\Delta s)/\Delta s$ and applied at the midpoint of the thrust arc.

With spin-stabilised spacecraft (Figure G) the thrust pulses are often fired relative to the Sun direction with a delay angle that is constant for the total thrust duration. The thrust direction will be constant in inertial space and is set to be tangential to the orbit at the mid-point of the thrust. Since it deviates from the tangential direction everywhere else its effect on the semimajor axis and longitude drift rate is reduced by the same factor as above, namely $2\sin(0.5\Delta s)/\Delta s$. In this case the eccentricity vector change is more complicated because of the varying tangential and radial components of the acceleration.

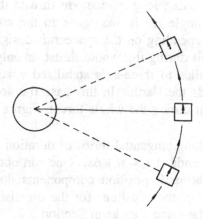

Figure 3.3.F. A long tangential thrust with a three-axis stabilised spacecraft.

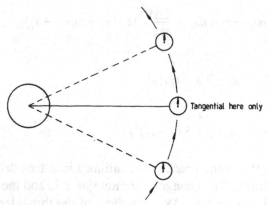

Tangential here only

Figure 3.3.G. A long thrust that is tangential only at one instant, with a spin-stabilised spacecraft.

For completeness, we also give here the equations that describe the effect of a radial thrust, although it is seldom used in practice. The ΔV_r is counted positive up, out from the Earth. The equations for the spacecraft position are the same as the first two equations in this section. The remaining two, for the velocity, become according to Section 3.1:

$$\Delta V_r = V\,(\Delta e_x\,\sin s_b - \Delta e_y\,\cos s_b)$$

$$0 = V(\Delta D + 2\Delta e_x\,\cos s_b + 2\Delta e_y\,\sin s_b)$$

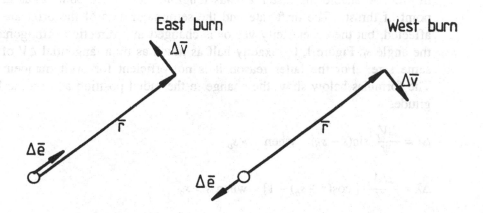

Figure 3.3.H. Effect of a tangential thrust on the eccentricity vector. $\Delta V = 1$ m/s gives $|\Delta\bar{e}| = 0.000650$

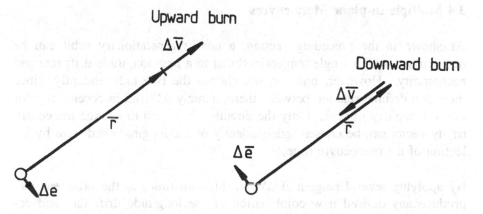

Figure 3.3.I. Effect of a radial thrust on the eccentricity vector. $\Delta V = 1$ m/s gives $|\Delta\bar{e}| = 0.000325$

The solution becomes

$$\Delta\lambda_0 = -2\Delta V_r/V$$

$$\Delta D = 0$$

$$\Delta\bar{e} = \frac{\Delta V_r}{V}\begin{pmatrix}\sin s_b \\ -\cos s_b\end{pmatrix}$$

Also here the mean longitude at epoch is changed for mathematical reasons in order to enable the instantaneous longitude to remain continuous at the point of thrust. The drift rate and the semimajor axis of the orbit are not affected, but the eccentricity vector is changed in a direction orthogonal to the angle s_b, Figure I, by exactly half as much as by a tangential ΔV of the same size. For the latter reason it is not efficient for orbit manoeuvres. The formulas below show the change in the radial position and on the longitude.

$$\Delta r = \frac{\Delta V_r}{\psi}\sin(s - s_b) \quad \text{when } s > s_b$$

$$\Delta\lambda = \frac{2\Delta V_r}{V}[\cos(s - s_b) - 1] \quad \text{when } s > s_b$$

3.4 Multiple In-plane Manoeuvres

As shown in the preceding section, a nearly geostationary orbit can be manoeuvred with a single tangential thrust to a new longitude drift rate and eccentricity. However, one can not choose the two independently, since there is a definite relation between them, namely 0.00185 in eccentricity for every 1 deg/day in drift. Only the direction in which to change the eccentricity vector can be chosen independently of the longitude drift rate by selection of the manoeuvre time.

By applying several tangential ΔVs at different times to the orbit one can produce any desired new combination of the longitude drift rate and eccentricity, independently of each other. The new orbital elements can, with good linear approximation, be obtained by the superposition of the drift rates (with the appropriate sign) and eccentricity vectors (by vector addition) from each individual thrust.

Without comparison, the most often used multiple in-plane thrust sequence is the two-burn longitude manoeuvre. It makes use of two ΔVs separated by half an orbital revolution, i.e. half a sidereal day = 11 hours and 58 minutes. It has got three free parameters, of which two are the sizes of the two ΔV. The third parameter is the time or spacecraft sidereal angle at the execution of one of the thrusts, whereas the time of the second thrust is fixed to half a day before or after.

We denote the two velocity increments by ΔV_1 and ΔV_2 and the corresponding spacecraft sidereal angles at the execution by s_1 and $s_2 = s_1 + \pi$, respectively. The combined effect on the semimajor axis, longitude drift rate and eccentricity vector is obtained by adding two ΔVs from the equations of the preceding section.

$$\Delta a = \frac{2}{\psi}(\Delta V_1 + \Delta V_2)$$

$$\Delta D = -\frac{3}{V}(\Delta V_1 + \Delta V_2)$$

$$\Delta \bar{e} = \frac{2}{V}(\Delta V_1 - \Delta V_2)\binom{\cos s_1}{\sin s_1} = \frac{2}{V}(\Delta V_2 - \Delta V_1)\binom{\cos s_2}{\sin s_2}$$

It is easy to see that by selecting ΔV_1, ΔV_2 and s_1 one can obtain any desired combination of ΔD and $\Delta \bar{e}$. In the time interval between the two thrusts the orbit changes under the influence of ΔV_1 alone according to the equations of the single thrust of the previous section. After the second thrust, when $s \geq s_2$, the in-plane orbital motion is obtained as before by adding the effect of two ΔVs on the orbit evolution. The result can be expressed as ($s > s_2$):

$$\Delta r = \frac{2}{\psi}[\Delta V_1 + \Delta V_2 - (\Delta V_1 - \Delta V_2)\cos(s - s_1)]$$

$$\Delta \lambda = \frac{1}{V}(\Delta V_1 - \Delta V_2)[4\sin(s - s_1) - 1.5\pi] - \frac{3}{V}(\Delta V_1 + \Delta V_2)(s - s_1 - \pi/2)$$

If one wants to change the drift rate without changing the eccentricity one shall perform the two-burn longitude manoeuvre using two thrust with the same ΔV. The effect on the eccentricity vector cancels, whereas the drift rate is added. Two positive ΔVs applied to an initially geostationary orbit will produce a higher circular orbit with negative longitude drift, as shown

in Figure A. The intermediate orbit, between the two thrusts, is the same
as in Figure 3.3.C. The longitude evolution is as in Figure 3.3.D for the first
half revolution, after which it becomes a straight line with twice the drift
rate. The sequence of two tangential thrusts to go from one circular orbit
to another shown in Figure A is known as the classical Hohmann transfer.

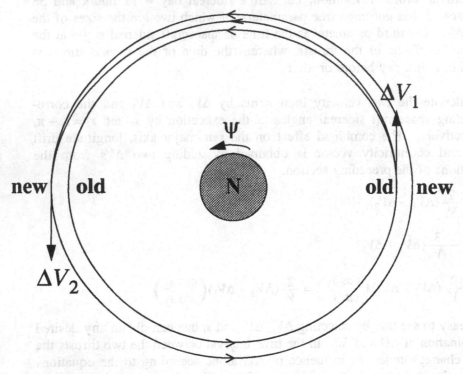

Figure 3.4.A. Two east thrusts to raise a circular orbit without changing the
eccentricity.

If one needs to change the eccentricity but not the longitude drift rate one
can use two thrusts of equal size but opposite signs, Figure B. The first,
east, thrust is the same as in Figures A and 3.3.C, but the second, west,
thrust now decreases the orbit back to the original semimajor axis. The
corresponding longitude evolution is shown in Figure C. The longitude of
the new orbit librates around a mean longitude that is shifted west from the
original one by $-3\pi\Delta V_1/V$, when $\Delta V_2 = -\Delta V_1$. Essentially the same effect
is obtained by performing the west thrust first and the east thrust half an
orbit later. The mean longitude is shifted as, before, but to the east instead

of west. In both cases the perigee of the new eccentric orbit will be positioned at the place where the east thrust was performed.

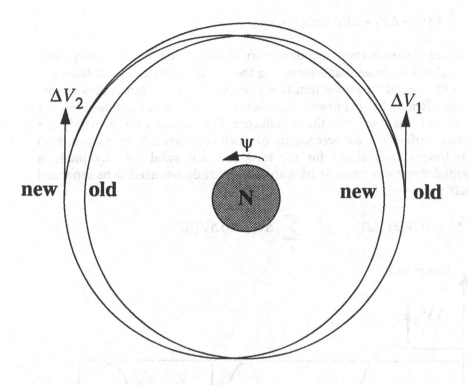

Figure 3.4.B. An east thrust followed by a west thrust to change the eccentricity of an initially circular orbit.

If one does not want the mean longitude to be shifted by the two-burn manoeuvre above there is the possibility to execute an east - west - east sequence as shown in Figures D and E, where the two east thrusts are of the same size and the west thrust is twice as strong. The resulting orbit is the same as of the two-burn pair of Figures B and C, except that the mean longitude of the new orbit is exactly the same as in the longitude before the manoeuvres. The result of the general three-burn manoeuvre is best expressed with the execution times symmetrically around the sidereal angle s_2 of the second thrust, with $s_1 = s_2 - \pi$, $s_3 = s_2 + \pi$ and $s > s_3$.

$$\Delta r = \frac{2}{\psi} [\Delta V_1 + \Delta V_2 + \Delta V_3 + (\Delta V_1 - \Delta V_2 + \Delta V_3) \cos(s - s_2)]$$

$$\Delta\lambda = \frac{3\pi}{V}(\Delta V_3 - \Delta V_1) - \frac{3}{V}(\Delta V_1 + \Delta V_2 + \Delta V_3)(s - s_2) -$$

$$- \frac{4}{V}(\Delta V_1 - \Delta V_2 + \Delta V_3)\sin(s - s_2)$$

The above mentioned two- and three-burn sequences are the most frequently used multiple in-plane manoeuvres. In theory, many possible combinations exist with several tangential thrusts separated by various time intervals, but they are often of limited practical importance. In order to optimise the use of fuel, one shall select a thrust sequence that depends on if mainly the longitude drift ΔD or the eccentricity $\Delta\bar{e}$ shall be corrected. In general, both the following inequalities for the total fuel are valid for any multiple tangential thrust sequence, at arbitrary times that do not need to be separated by half a sidereal day.

$$\sum_k |\Delta V_k| \geq (V/3)|\Delta D| \quad ; \quad \sum_k |\Delta V_k| \geq 0.5V|\Delta\bar{e}|$$

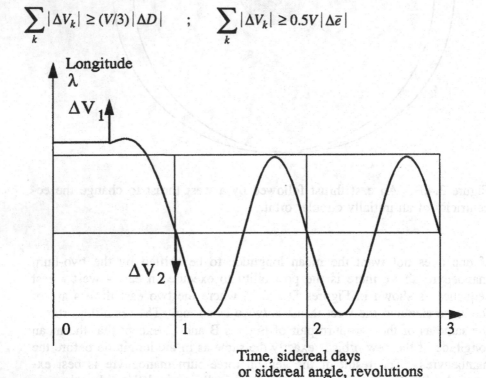

Figure 3.4.C. The longitude evolution resulting from the east-west thrust pair of Figure B.

The fuel is optimised for a desired change of the orbit if at least one of the inequalities above becomes an equality. The result depends on the condition if the desired drift change $|\Delta D|$ is greater or smaller than 1.5 times the desired change in eccentricity $|\Delta \bar{e}|$, as follows.

- If the change in the mean longitude drift rate is most important, $|\Delta D| \geq 1.5 |\Delta \bar{e}|$, all the thrusts must be in the same direction but they do not necessarily need to be separated by half a sidereal day.

- If the change in the eccentricity is most important, $|\Delta \bar{e}| \geq |\Delta D|/1.5$, one must perform a sequence of alternating east and west thrusts, separated by half a sidereal day.

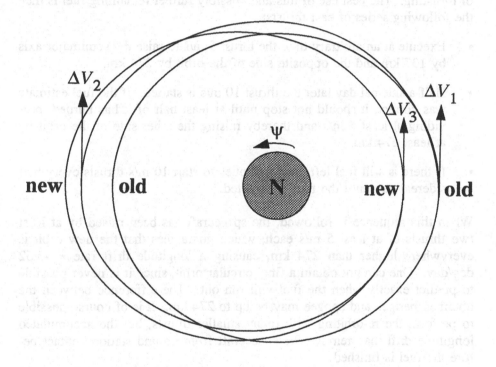

Figure 3.4.D. An east-west-east thrust sequence to change the eccentricity of an initially circular orbit.

A series of tangential thrusts is recommended for removing each old spacecraft from the geostationary orbit at its end of life by raising it to another circular orbit above the geostationary region. The exact amount to raise it is the subject of different opinions, but it should be of the order of a few 100 km. For planning such re-orbiting manoeuvres, one must first try to estimate how much fuel is left. This is often difficult to do accurately, and only when a thruster stops during firing does one know with certainty that the fuel is finished.

For the numerical example below we assume that the end of the mission is approaching but that the estimated fuel left corresponds to at least 10 m/s of thrusting. The best use of this and possibly further remaining fuel is then the following series of east thrusts:

- Execute at an arbitrary time the thrust 5 m/s to raise the semimajor axis by 137 km and the opposite side of the orbit by 274 km.

- Half a sidereal day later the thrust 10 m/s is started. If the fuel estimate was correct, it should not stop until at least half of it has burned, providing at least 5 m/s and thereby raising the other side of the orbit by at least 274 km.

- If there is still fuel left, one continues to start 10 m/s thrusts every half sidereal day until the fuel is depleted.

When this sequence is followed, the spacecraft has been raised by at least two thrusts of at least 5 m/s each, which guarantees that the new orbit is everywhere higher than 274 km, causing a longitude drift rate = -3.52 deg/day. One can not obtain a final circular orbit, since it is never possible to predict exactly when the fuel will run out. The difference between the obtained perigee and apogee may be up to 274 km. It is of course possible to perform the re-orbiting with many smaller thrusts, but the accumulated longitude drift may remove the spacecraft from ground station contact before the fuel is finished.

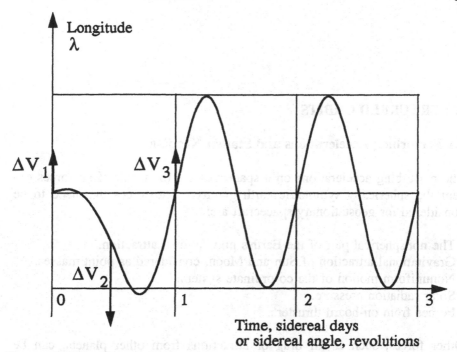

Figure 3.4.E. The longitude evolution resulting from the east-west-east thrust sequence of Figure D.

4. PERTURBED ORBITS

4.1 Perturbing Accelerations and Station Keeping

The perturbing accelerations on a spacecraft are the result of all forces except the spherically symmetric earth gravity. The forces that need to be considered for geostationary spacecraft are:

- The nonspherical part of the Earth's gravitational attraction.
- Gravitational attraction of Sun and Moon, considered as point masses.
- Nonuniform motion of the co-ordinate system.
- Solar radiation pressure.
- Forces from on-board thrusters.

Other forces, such as air drag or attractions from other planets, can be neglected. The exact spacecraft motion can be expressed by the differential equation of the spacecraft position $\bar{r}$ or of the state vector $(\bar{r}, \bar{V})$

$$\frac{d^2\bar{r}}{dt^2} = -\frac{\mu}{r^3}\bar{r} + \text{ sum of perturbations}$$

$$\frac{d}{dt}\left(\frac{\bar{r}}{\bar{V}}\right) = \left(\begin{array}{c} \bar{V} \\ -(\mu/r^3)\bar{r} + \text{ sum of perturbations} \end{array}\right)$$

The perturbing accelerations can be expressed with high accuracy as functions of the spacecraft position $\bar{r}$ and the time t, since the position of Sun and Moon and the orientation of the Earth are known as functions of time. The equation for the state vector can be integrated numerically with respect to time by, say, a standard eighth-order multistep method in double precision on a modern computer. No other conversions or transformations are needed, although many are proposed in celestial mechanics text.

In order to show qualitatively the effect of the perturbations on the orbit, we will continue the linearisation of the spacecraft motion begun in Section 2.3. In this way, one can describe the spacecraft motion in the earth-rotating frame as a sum of contributions from the perturbations and from the deviations of the orbital elements from the true geostationary values.

The size of the spacecraft motion due to the perturbations is approximately independent of the orbital elements and depends on the source of the perturbations. Details will be given in the following sections.

Since the mean synchronous elements are averages over one sidereal day, it is convenient to split the effect of perturbations into two parts:

- short-term perturbations that are periodic with a period of one day or shorter;

- long-term perturbations that change the mean orbital elements. The long-term perturbations can be further divided into parts with periods from a few weeks up to several years.

The purpose of the orbital station keeping manoeuvres is to compensate for the effect of the long-term perturbations and to keep the orbital elements within the prescribed boundary. The longitude and latitude region within which the spacecraft must be kept is called the *deadband* or control box, and its size depends on the requirements of the mission. The smaller the deadband, the more frequently the station keeping manoeuvres must be performed. It is not realistic to compensate for the short-term perturbations in the station keeping, so there is a lower limit to the size of the deadband for the spacecraft position in the earth-rotating co-ordinate system.

Certain perturbing effects dominate the long-term evolution of certain orbital elements. In a first-order approximation we have:

- The mean longitude drift rate D is perturbed practically only by the tesseral terms of the Earth's gravity;

- The eccentricity vector is perturbed mainly by the solar radiation pressure, and slightly by solar and lunar attraction;

- The inclination vector drift is dominated by the attraction of the Sun and Moon with contribution from the zonal terms of the Earth's gravity.

Among the perturbing forces one must also count thruster burns on board the spacecraft. The thrusts are not necessarily intended for orbit control. Depending on the spacecraft design, there may be thruster burns executed to keep or change the orientation (= attitude) of the spacecraft around its centre of gravity. If these thrusts are performed by an automatic on-board control system it may be difficult to take them into account in the differential equation that represents the orbit evolution on a ground computer.

The perturbing forces from the gravitational attractions of the Sun, Moon and Earth are all proportional to the mass of the spacecraft. One divides by the mass to obtain the spacecraft acceleration, so these perturbing accelerations are independent of the size, mass, shape or any other properties of the spacecraft. On the other hand, the perturbing forces from the solar radiation pressure and from manoeuvre burns do not depend on the mass, so the accelerations become inversely proportional to the spacecraft mass.

4.2 Nonspherical Earth Potential and Longitude Evolution

The gravitational attraction of the Earth is not sufficiently symmetric to be considered as from a point mass or a sphere for modelling the motion of an Earth-orbiting spacecraft. It is normal to express the acceleration on a spacecraft from a gravity field as the gradient of a potential function U. (In some physics texts, potential functions are defined with the opposite sign.)

$$\text{Acceleration} = \text{grad } U = (\frac{\partial U}{\partial x}, \frac{\partial U}{\partial y}, \frac{\partial U}{\partial z})$$

The potential function is defined by the coefficients of the multipole expansion (series development in spherical harmonics) in the Earth-rotating system. Different sets of coefficients with varying number of terms are available in publications from geodetic institutes. Mostly used are the gravitational models from Goddard Space Flight Center (USA). The coefficients are dimension-less since they are multiplied by μ and powers of the Earth's equatorial radius R.

$$U = \frac{\mu}{r} + \mu \sum_{l=2}^{L} \sum_{m=0}^{l} \frac{R^l}{r^{l+1}} P_{lm} (\sin \theta) (C_{lm} \cos m\lambda + S_{lm} \sin m\lambda)$$

The associated Legendre functions of degree l and order m are here denoted by P_{lm}. For geostationary orbits an expansion to order $L = 8$ is in most cases sufficiently accurate. The term with $l = m = 0$ is the central Earth potential μ/r. The three coefficients with $l = 1$ (C_{10}, C_{11}, S_{11}) are zero by definition, because the origin of the co-ordinate system is defined to be located at the mass-centre of the Earth. Both statements are equivalent to the fact that all three components are zero of the following integral over the whole mass of the Earth:

$$\int \bar{r} dm = 0$$

The zonal terms are the C terms with $m = 0$ since no S terms with $m = 0$ exist. (They would have been multiplied by zero.) The zonal part of the potential is rotationally symmetric, i.e. independent of λ, and is mainly caused by the Earth's flatness. Its effect on a geostationary spacecraft is an acceleration of:

- in tangential direction $= 0$
- in north direction $= -2.95 \times 10^{-9} \, \text{m/s}^2$
- in radial direction $= -8.33 \times 10^{-6} \, \text{m/s}^2$

The radial component has the same effect on the orbit as the central potential and increases the value of the true geostationary radius by 0.5 km. The north component influences the direction of the inclination vector drift.

The tesseral terms are the terms where $m \geq 1$ in the expansion of U and are caused by the unsymmetrical mass distribution inside the Earth. The tangential component of the acceleration is of great importance for the longitude drift evolution in spite of the fact that it is smaller that $10^{-7} \, \text{m/s}^2$ in absolute value. It is tabulated (B) as a function of the spacecraft longitude in Table 2. The radial and north components have effects similar to those of the zonal terms, only smaller.

Because the tangential acceleration is dominated by the two coefficients with $l = m = 2$ its longitude dependence becomes approximately sinusoidal with four nodes, as shown in Figure A. At the four nodes around the Earth where the acceleration is zero, a geostationary spacecraft can stay at rest with respect to longitude. Two of these are stable equilibrium points, since any small longitude deviation from the node would induce a drift back towards the node, Figure A. The other two nodes are unstable, from which a spacecraft will drift away in either direction.

- Stable points $= 75.1° \, \text{E}$ and $105.3° \, \text{W}$
- Unstable points $= 11.5° \, \text{W}$ and $161.9° \, \text{E}$

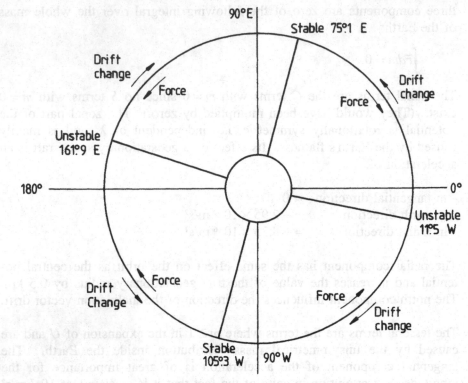

Figure 4.2.A. Change in longitude drift due to the Earth's gravity field.

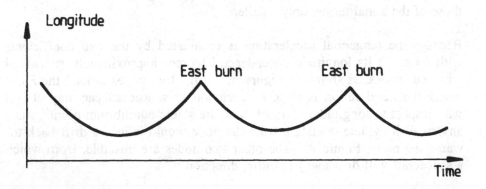

Figure 4.2.B. Parabolic shape of the mean longitude free drift around 0°
longitude with station keeping by east thrusts.

When a geostationary spacecraft is kept by longitude manoeuvres inside a narrow deadband, say less than 1°, the tesseral acceleration can be considered to be constant. Its tangential component (B) will cause a change of the orbit semimajor axis and mean longitude drift rate, which is the same as the effect of a weak continuous longitude thrust. Because of the rotational symmetry, the orbit eccentricity does not change, but the drift rate changes continuously in a way that is analogous to the change by a manoeuvre ΔV:

$$\frac{d^2\lambda}{dt^2} = -\frac{3}{A} B$$

In this context we use $d^2\lambda/dt^2$ instead of dD/dt to describe the drift rate changes since the orbital element D is not suitable to describe the obtained parabolic drift of λ. Table 2 also provides the corresponding longitude change $d^2\lambda/dt^2$ converted to degrees per day squared. If we take B to be constant we can integrate the differential equation above twice to obtain the mean longitude evolution. If the longitude is not near one of the four nodes, the small longitude variation as a function of time describes a series of parabolas joined by cusps at the station keeping thrusts, Figure B. The curvature of the parabolas may be turned either way, depending on the sign of the acceleration, i.e. depending on the mean longitude.

The last column of Table 2 gives the average fuel consumption for a mission expressed as m/s per year for longitude station keeping. This is an accurate way of calculating the fuel when the orbit eccentricity stays so small as not to require special correction manoeuvres. On the other hand, spacecraft with high solar radiation perturbations often require additional fuel for east-west manoeuvres in order to reduce the buildup of the eccentricity.

A geostationary spacecraft at any longitude is accelerated towards the nearest of the two stable equilibrium points, Figure A. If no station keeping manoeuvres are executed, its longitude will eventually swing back and forth symmetrically around the stable point like a pendulum with a period of more than two years. Since there is no damping the swing will continue indefinitely with approximately constant amplitude. A spacecraft can be at rest at a stable point only if a longitude drift stop manoeuvre is performed when it arrives at this longitude. Once this is done, however, it can stay at this longitude for ever without station keeping manoeuvres. It will only experience librations in longitude due to eccentricity buildup and an increasing orbital inclination from solar and lunar perturbations.

Below follows a short account of how the spherical harmonics and their gradients can be calculated. The method is general and can be used for all types of orbits, not only for geostationary missions. It is quite efficient since it does not require the computation of trigonometric functions inside the inner loops, and it has no singularities.

The Legendre polynomials P_l of degree l of the independent variable $u = \sin \theta = z/r$ are defined by:

$$P_l(u) = \frac{1}{2^l l!} \frac{d^l}{du^l} (u^2 - 1)^l$$

They are best calculated, for a given input value of u, by the following recursive formula with the starting values:

$$P_0(u) = 1 \quad ; \qquad P_1(u) = u$$

$$P_l(u) = \frac{2l-1}{l} u P_{l-1}(u) - \frac{l-1}{l} P_{l-2}(u) \quad \text{when } l \geq 2$$

The associated Legendre function for a given order m and degree l is defined by:

$$P_{lm}(u) = (1 - u^2)^{m/2} \frac{d^m P_l(u)}{du^m}$$

With these definitions the spherical harmonics do not become normalised. In order to normalise them such that the integral over the sphere, in longitude and latitude, of the square of each function becomes $= 4\pi$ one needs to multiply by

$$\sqrt{(2l+1)} \quad \text{when } m = 0 \quad ; \qquad \sqrt{2(2l+1)\frac{(l-m)!}{(l+m)!}} \quad \text{when } m \geq 1$$

Instead of multiplying these normalisation factors at each computation step it is more efficient to multiply the corresponding zonal and tesseral coefficients by them only once, before starting the computation. The resulting first six coefficients are listed below.

$$C_{20} = -1.0826 \times 10^{-3} \quad C_{21} = 0 \qquad\qquad\qquad S_{21} = 0$$
$$C_{22} = +1.57 \times 10^{-6} \quad\, S_{22} = -0.90 \times 10^{-6} \qquad C_{30} = +2.53 \times 10^{-6}$$

It is, for computational reasons, a great advantage to replace the associated Legendre functions of order m by the m-th derivative of the corresponding

Legendre polynomials, i.e. we move the factor $(1 - u^2)^{m/2} = \cos^m\theta$ to the sine and cosine functions. The advantage is due to the fact that it simplifies the calculation of the derivatives that are needed for the gradient of U. It also removes the singularities that would otherwise appear at the latitudes $\pm 90°$, although this is not of practical interest for geostationary orbits. We denote:

$$P_l^{(m)}(u) = \frac{d^m P_l(u)}{du^m}$$

which can be calculated by the following recursive formula in l for given values of u and $m \geq 1$ with the starting values:

$$P_l^{(m)}(u) = 0 \quad \text{when } l < m$$

$$P_l^{(m)}(u) = 1 \times 3 \times \ldots \times (2m - 1) \quad \text{when } l = m$$

$$P_l^{(m)}(u) = \frac{2l - 1}{l - m} u P_{l-1}^{(m)}(u) - \frac{l + m - 1}{l - m} P_{l-2}^{(m)}(u) \quad \text{when } l > m$$

For the subsequent calculations we introduce two components of the spacecraft position vector in the Earth-rotating coordinate system, by means of the Greenwich sidereal angle G.

$$x_E = r \cos \theta \cos \lambda = x \cos G + y \sin G$$

$$y_E = r \cos \theta \sin \lambda = y \cos G - x \sin G$$

The λ-dependent part of the spherical harmonics can now be expressed by the introduction of the two parameters ξ_m and η_m. The following two definitions are equivalent, but the second one, which expresses them as the real and imaginary parts of a complex parameter, is easier to visualise. We denote here the imaginary unit by j ($j^2 = -1$) so as to distinguish it from the inclination.

$$\xi_m = r^m \cos^m\theta \cos m\lambda \quad ; \quad \eta_m = r^m \cos^m\theta \sin m\lambda$$

$$\xi_m + j\eta_m = (r \cos \theta e^{j\lambda})^m = (x_E + jy_E)^m$$

They can be calculated by the recursive formulas:

$$\xi_m = \xi_{m-1} x_E - \eta_{m-1} y_E \quad \text{starting with } \xi_0 = 1$$

$$\eta_m = \xi_{m-1} y_E + \eta_{m-1} x_E \quad \text{starting with } \eta_0 = 0$$

By inserting the new functions into each term of the multipole expansion at the beginning of this section, we obtain its gradient by a simple calculation:

$$\text{grad} \left[\frac{\mu R^l}{r^{l+m+1}} P_l^{(m)}\left(\frac{z}{r}\right) (C_{lm}\,\xi_m + S_{lm}\,\eta_m) \right] =$$

$$= -(l+m+1) \frac{\mu R^l}{r^{l+m+3}} P_l^{(m)}\left(\frac{z}{r}\right) (C_{lm}\,\xi_m + S_{lm}\,\eta_m)\,\bar{r} +$$

$$+ \frac{\mu R^l}{r^{l+m+1}} (C_{lm}\,\xi_m + S_{lm}\,\eta_m)\,\text{grad}\,P_l^{(m)}\left(\frac{z}{r}\right) +$$

$$+ \frac{\mu R^l}{r^{l+m+1}} P_l^{(m)}\left(\frac{z}{r}\right) (C_{lm}\,\text{grad}\,\xi_m + S_{lm}\,\text{grad}\,\eta_m)$$

Here we insert the gradients of the functions that were defined above:

$$\text{grad}\,P_l^{(m)}(z/r) = P_l^{(m+1)}(z/r)\,[(1/r)(0,0,1) - (z/r^3)\bar{r}]$$

$$\text{grad}\,\xi_m = m(\xi_{m-1}\cos G + \eta_{m-1}\sin G,\; \xi_{m-1}\sin G - \eta_{m-1}\cos G,\; 0)$$

$$\text{grad}\,\eta_m = m(-\xi_{m-1}\sin G + \eta_{m-1}\cos G,\; \xi_{m-1}\cos G + \eta_{m-1}\sin G,\; 0)$$

4.3 Solar and Lunar Attraction in a Moving System

The gravitational attraction of the Sun and the Moon, considered as point masses, on the spacecraft satisfies the same inverse-square law as the attraction of the Earth. Because of the great distance, there is no need to take the nonspherical part of the gravity into consideration. For the computation we need to express in the computer the positions of the Sun and Moon in our co-ordinate system MEGSD as functions of time. This can be done by transformation of one of the numerical planetary ephemerides published by Jet Propulsion Laboratory (JPL) in the USA or by using an analytical ephemeris from current literature.

We denote the gravity potentials of the Sun and Moon by μ_1 and μ_2, in an analogous way as μ for the Earth, and their position vectors in the MEGSD coordinate system by $\bar{r}_1$ and $\bar{r}_2$. The position vectors of the Earth, Sun and Moon relative to the spacecraft are, respectively, $-\bar{r}$, $\bar{r}_1 - \bar{r}$ and $\bar{r}_2 - \bar{r}$. Their combined gravitational attraction on the spacecraft in a coordinate system that is momentarily at rest, is =

$$= -\frac{\mu}{r^3}\bar{r} + \sum_{k=1}^{2} \frac{\mu_k}{|\bar{r}_k - \bar{r}|^3} (\bar{r}_k - \bar{r})$$

However, also the Earth moves under the attraction of the Sun and Moon. The acceleration of the Earth is the same as the acceleration of the MEGSD coordinate system, in which $\bar{r}$, $\bar{r}_1$ and $\bar{r}_2$ are expressed. The acceleration of the Earth is =

$$= \sum_{k=1}^{2} \frac{\mu_k}{r_k^3} \bar{r}_k$$

which shall be subtracted on the right hand side of the previous equation. The result is the acceleration and subsequent motion of the spacecraft in MEGSD

$$\frac{d^2\bar{r}}{dt^2} = -\frac{\mu}{r^3}\bar{r} + \sum_{k=1}^{2} \mu_k \left[\frac{\bar{r}_k - \bar{r}}{|\bar{r}_k - \bar{r}|^3} - \frac{\bar{r}_k}{r_k^3} \right]$$

Since r_k is much greater than r, the two terms inside the summation bracket almost cancel out. To better visualise the remaining effect one can perform a linear expansion in r/r_k, although this approximation is not necessary when the perturbed orbit is integrated by numerical means. The following is the result of linearising the content of the bracket:

$$\frac{d^2\bar{r}}{dt^2} \approx -\frac{\mu}{r^3}\bar{r} + \sum_{k=1}^{2} \frac{\mu_k}{r_k^3} \left[\frac{3}{r_k^2} (\bar{r}_k \cdot \bar{r})\bar{r}_k - \bar{r} \right]$$

The last term inside the bracket is a negative, almost constant, radial acceleration on the spacecraft. The preceding term is the acceleration caused by the perturbing body, i.e. the Sun or Moon, respectively, in the direction to the body. However, the term is positive only during half the day when the spacecraft is at the same side of the Earth as the body and thus is stronger attracted, Figure A. The other half of the day its distance is greater than that of the Earth and thus is less attracted, so the term is negative. This results in a variable net acceleration on the spacecraft, in MEGSD, in the positive radial direction on either side of the Earth, Figure B. The term becomes zero when the Earth and the body are seen at right angles from the spacecraft. There are also time-varying accelerations in the tangential and out-of-plane directions.

The following table shows the size and characteristics of the perturbations. We see that the high gravity potential of the Sun is more than compensated for by its distance, compared with that of the Moon. The net resulting perturbation from the Moon is then about twice as strong as from the Sun.

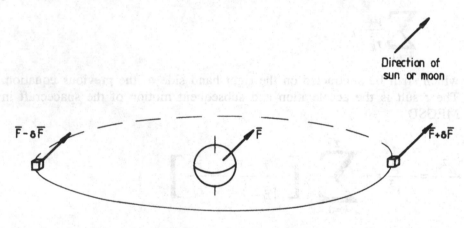

Figure 4.3.A. Solar or lunar gravity attraction $\bar{F} \pm \delta \bar{F}$ in the inertial system.

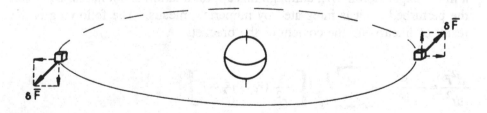

Figure 4.3.B. Net solar or lunar gravity attraction $\delta \bar{F}$ in the mean equatorial geocentric system of date (MEGSD).

Perturbing Body =	*Sun*	*Moon*
Period of motion =	1 year	27 days
Maximum declination in MEGSD =	23.442°	28.6°
Gravity potential = μ_k km³/s² =	1.327×10^{11}	4.903×10^3
Mean distance from Earth = r_k km =	149.6×10^6	385.0×10^3
Acceleration in inertial system = μ_k/r_k^2 m/s² =	5.93×10^{-3}	3.31×10^{-5}
Acceleration in MEGSD = $3\mu_k r/r_k^3$ m/s² =	0.5×10^{-5}	1.1×10^{-5}

In principle, the same type of attraction on the spacecraft and on the Earth is exerted by all celestial bodies within or outside the planetary system. However, because the net attraction in MEGSD decreases with the cube of the distance, the only noticeable attractions are obtained from bodies that are either close but small (Moon) or heavy but distant (Sun). Of other bodies, Venus causes the highest acceleration in MEGSD of 6×10^{-10} m/s² when it is close to the Earth.

In addition to the translational motion of MEGSD, one must also consider the acceleration from the precession of the co-ordinate system, as described in Section 2.1. The Coriolis acceleration is of the order of 4×10^{-8} m/s² for a geostationary orbit, which is added as a perturbation on the right-hand side of the differential equation of spacecraft motion in MEGSD.

We can qualitatively describe the effect of all these perturbations on the geostationary orbit by splitting the net acceleration of the spacecraft in MEGSD into an in-plane and an out-of-plane component resulting in an in-plane and an out-of-plane effect on the orbit, respectively, Figure B. The out-of-plane effect is important for the long-term evolution of the inclination vector and is dealt with in the next section.

The positive radial acceleration attains its maximum twice per day: when the spacecraft is at the same and on the opposite side of the Earth relative to the Moon or Sun, Figure B. This perturbing effect on the spacecraft orbit is strongest when the Sun, Moon and Earth are closely aligned twice per month, Figure C. The same effect is noticed on the Earth as the spring tides of the ocean surface. The smallest perturbation is obtained when the Sun and Moon are seen at right angles from the Earth, Figure C. The corresponding ocean tide is called a neap tide.

Figures D and E show plots of the short term variation of the in-plane spacecraft position and osculating orbital elements for orbits that are chosen

to be as close to geostationary as possible during one sidereal day. Plotted
are the spacecraft distance from the Earth r and the longitude λ together
with the semimajor axis a and the two eccentricity vector components e_x and
e_y. Figure D shows the orbit for a spring tide and Figure E for a neap tide.
The highest in-plane effect is obtained when the Sun and Moon lie in the
orbital plane, which here is the same as the equatorial plane. When the
declination of the Sun or Moon is non-zero the corresponding perturbation
decreases by the factor cosine of the declination.

The perturbations from the zonal and tesseral terms of the Earth's gravity,
but not the solar radiation pressure, are included in the orbital motion that
is plotted in the figures. The tesseral terms at longitude zero cause the small
decrease in r and a, which is equivalent to the longitude acceleration, during
the sidereal day. There is a complicated cross-coupling between the
gravitational perturbations from the Earth, Sun and Moon on the combined
librations of r and λ so it is not possible to identify which effect is caused
by which perturbation.

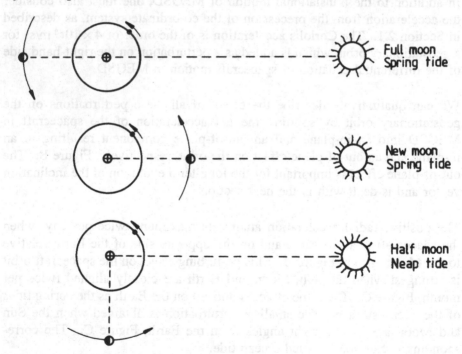

Figure 4.3.C. Tidal effect of Sun and Moon.

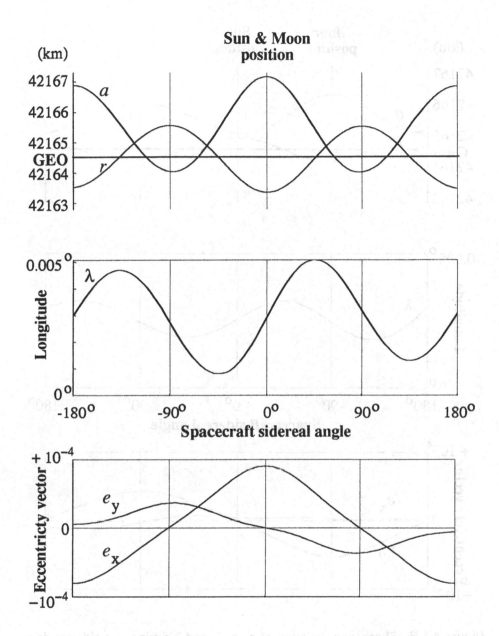

Figure 4.3.D. Short-term variation of r, a, λ and $\bar{e}$ during one sidereal day at spring tide: the Sun and the Moon are in the +x-direction.

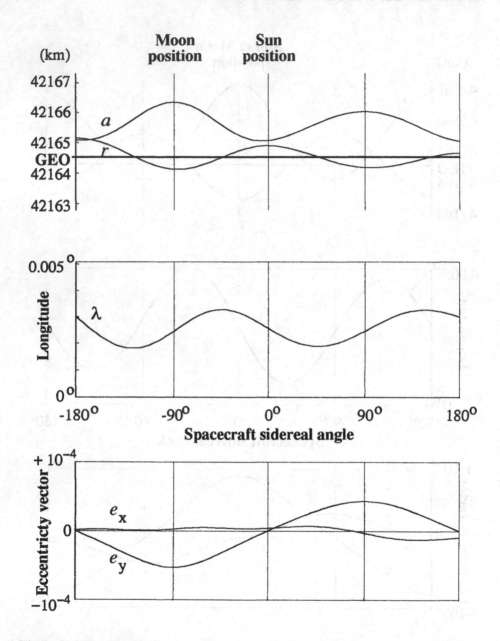

Figure 4.3.E. Short-term variation of r, a, λ and $\bar{e}$ during one sidereal day at neap tide: the Sun is in the +x-direction and the Moon is in the -y-direction.

We see in Figures D and E that r, a and λ librate with a period of half a sidereal day, whereas the motion of $\bar{e}$ is equivalent to a negative rotation of its direction with the period of one sidereal day. The spacecraft motion, represented by the libration of r and λ, is a real physical effect, whereas the motion of the osculating elements a and $\bar{e}$ is only a mathematical artefact. It follows from fitting orbital elements, which were originally defined for an unperturbed orbit, to be osculating elements of a perturbed orbit. However, the mean eccentricity vector, obtained from the mean value over one sidereal day of the two components, is meaningful for representing the mean orbit.

The short-term variations in r and λ provide the lower limit as to how close to geostationary an orbit can be. The highest perturbations, shown in Figure D during a spring tide, consist of a total variation in r of 2.2 km and in λ of $0.0041° = 3.0$ km in the along-track position. The geostationary radius is defined to be the mean value of the spacecraft distance from the centre of the Earth, namely 42164.5 km. This is smaller than the mean value of the geostationary semimajor axis a, which is 42165.8 km, Figures D and E. It is natural that these two mean values are different, considering that the latter is calculated by means of only the Earth's μ, whereas the former includes the effect of the complete Earth potential, as well as μ_* from the Sun and the Moon. Both mean values vary slightly in time from the time-dependent positions of the Sun and the Moon.

At spring tide the orbit has the shape of an ellipse with the minor axis aligned with the direction to the Sun and Moon. The Earth lies now at the centre of this ellipse and not at a focal point, in contrast to the situation for unperturbed motion. For that reason there is no direct correspondence between the actual eccentricity of the ellipse, which is as high as 0.01, and the osculating eccentricities shown in Figure D. The reason is that the latter are obtained by fitting osculating perturbed state vectors to a focal point at the Earth's centre.

A medium term motion of the mean eccentricity vector is circular in the positive direction with the same period as the Moon's orbit around the Earth and with a radius $= 3.5 \times 10^{-5}$ as seen in Figure F. The analogous effect from the Sun is hidden by the effect of the solar radiation pressure, which also causes the loops in Figure F to be drawn into a spiral motion.

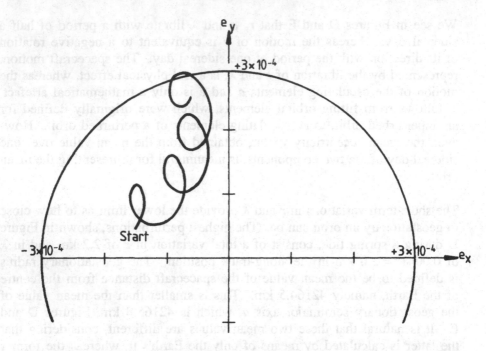

Figure 4.3.F. Motion of the mean eccentricity vector for ESA's Meteosat-1
from January to April 1980. The four small loops, one for each month, are
intermediate perturbation effects of the Moon's gravity.

4.4 Inclination Evolution

At midsummer and midwinter, when the Sun lies in the y-z-plane, the out-
of-plane component of the Sun's gravity perturbation in MEGSD gives a
north acceleration on the spacecraft during one half of the day and a south
acceleration during the other half, Figure A. The change from north to south
takes place when the y-component of the spacecraft position changes sign.
The effect on the spacecraft orbit is a torque that moves the inclination
vector in the +x-direction from the perturbations both at summer and winter.
In spring and autumn, when the Sun is aligned along the x-axis, the out-
of-plane perturbation is zero.

At other times of the year the Sun's effect on the orbit is similar: the inclination vector moves in a direction orthogonal to the sidereal angle of the Sun at a drift rate proportional to the out-of-plane perturbation torque. The result is a wavy motion with a semiannual period and a mean drift in the +x-direction of 0.27° per year.

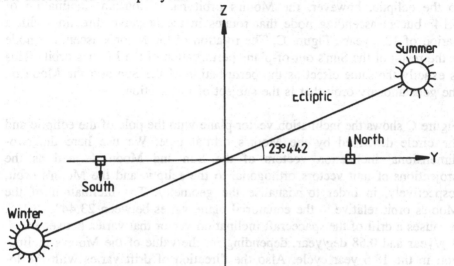

Figure 4.4.A. Out-of-plane component of the Sun's perturbation at midsummer and midwinter.

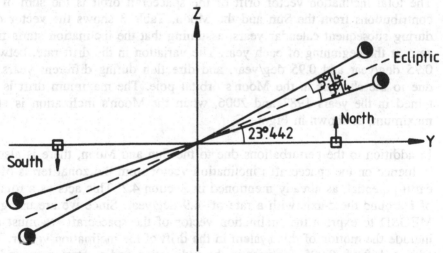

Figure 4.4.B. Out-of-plane component of the Moon's perturbation twice per month. The Moon's orbital plane moves with a period of 18.6 years.

The effect of the out-of-plane lunar perturbation on the orbital inclination is the same as that of the solar perturbation, Figure B. The acceleration is maximum twice per lunar period and passes through zero in between. The calculation of the lunar perturbation is complicated by the fact that the orbital plane of the Moon varies with respect to the x-y-plane. With respect to the ecliptic, however, the Moon's orbit has a constant inclination of 5.14° but an ascending node that rotates in the negative direction with a period of 18.6 years, Figure C. The rotation of the Moon's ascending node is the result of the Sun's out-of-plane perturbation of the Moon's orbit. This is exactly the same effect as the perturbation of the Sun and the Moon on the geostationary orbit that is the subject of this section.

Figure C shows the inclination vector plane with the pole of the ecliptic and the circle described by the Moon's orbital pole. We use here the two-dimensional inclination vectors of the Sun and Moon defined as the projections of unit vectors orthogonal to the ecliptic and the Moon's orbit, respectively, in order to visualise the geometry. The inclination of the Moon's orbit relative to the equatorial plane varies between 23.44° ± 5.14°. It causes a drift of the spacecraft inclination vector that varies between 0.48 deg/year and 0.68 deg/year, depending on the value of the Moon's inclination in the 18.6 year cycle. Also the direction of drift varies with the location of the Moon's pole.

The total inclination vector drift of the spacecraft orbit is the sum of the contributions from the Sun and the Moon. Table 3 shows the vector drift during subsequent calendar years, assuming that the inclination starts from zero at the beginning of each year. The variation in the drift rate, between 0.75 deg/year and 0.95 deg/year, and direction during different years are due to the changes in the Moon's orbital pole. The maximum drift is obtained in the years 1987 and 2006, when the Moon's inclination is at its maximum, as shown in Figure C.

In addition to the perturbations due to the Sun and Moon, there is also an influence on the spacecraft's inclination vector from the zonal terms of the earth potential, as already mentioned in Section 4.2. This acts as a rotation of $\bar{i}$ around the z-axis with a rate of -4.9 deg/year. Since we are using the MEGSD to express the inclination vector of the spacecraft we must also include the motion of this system in the drift of the inclination vector. This adds a drift of -0.005 deg/year in the x-direction and a rotation around the z-axis by +0.01 deg/year.

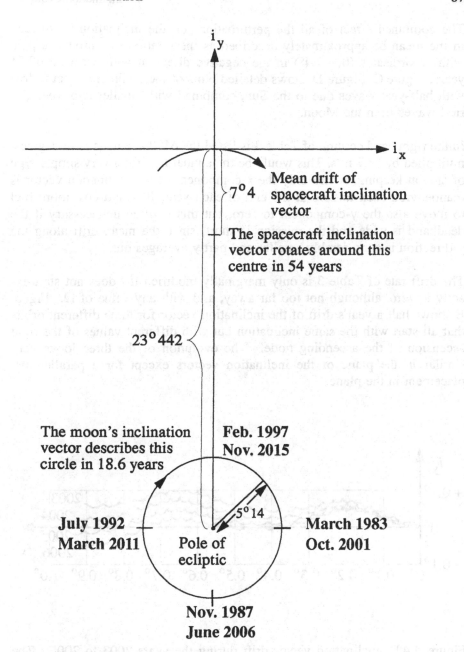

Figure 4.4.C. The two-dimensional inclination vectors of the spacecraft, Moon and Sun (ecliptic).

The combined effect of all the perturbations on the inclination vector can in the mean be approximately described as the rotation of $\bar{i}$ around a pole with co-ordinates $(0, -7.4°)$ in the negative direction with a period of 54 years, Figure C. Figure D shows detailed plots of the inclination vector drift with half-year waves due to the Sun, combined with smaller two-week period waves from the Moon.

In the right hand column of Table 3 is listed the ΔV that equals i_x, in degrees, multiplied by 53.7 m/s. This would be the annual fuel for a very simple type of station keeping, where only the x-component of the inclination vector is manoeuvred back to zero at the end of each year. It would cost more fuel to move also the y-component to zero, but this is often unnecessary if the deadband is wide and the mission is short, since the mean drift along the y-direction is considerably smaller and partly averages out.

The drift rate of Table 3 is only marginally modified if i does not start exactly at zero, although not too far away, and with any value of Ω. Figure E shows half a year's drift of the inclination vector for three different orbits that all start with the same inclination but with different values of the right ascension of the ascending node. The evolution of the three looks very similar in the plane of the inclination vectors except for a parallel displacement in the plane.

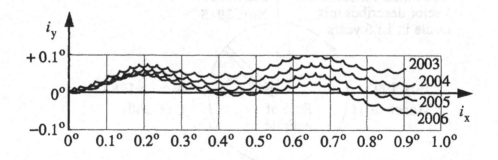

Figure 4.4.D. Inclination vector drift during the years 2003 to 2006. The inclination starts from $0°$ at the beginning of each year.

The plot of the same evolution in Figure F in the form of i and Ω as functions of time looks very different for the three orbits. The phenomenon that, depending on the initial position of the inclination vector, the inclination may or may not drift close to zero and the node may or may not jump from 360° to 0° is known to be a source of confusion to beginners in orbital mechanics. For this reason the type of plot as in Figure E is more illustrative for inclination drift and station keeping.

One consequence of the fact that the drift of the inclination vector only weakly depends on its starting position is that the total station keeping ΔV remains about the same regardless of how often the manoeuvres are performed. For this reason one can use Table 3 for a relatively accurate estimate of the annual ΔV for inclination station keeping. The required velocity increment fluctuates between 40 m/s and 51 m/s per year, which is very much more than is needed for longitude station keeping: up to 2 m/s per year according to Table 2. However, slightly higher or lower rates can be obtained depending upon the size of the deadband and the degree of sophistication of the optimisation program, Section 6.4.

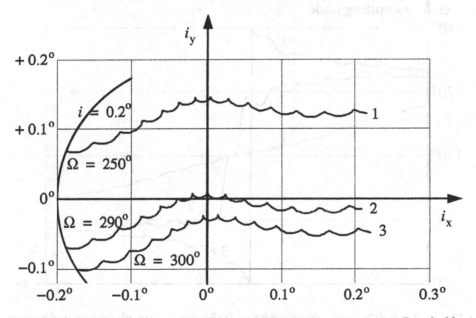

Figure 4.4.E. Inclination vector drift of three orbits during the first half of the year 2000, all with the initial $i = 0.2°$ but with the initial $\Omega = 250°$, 290°, 300°.

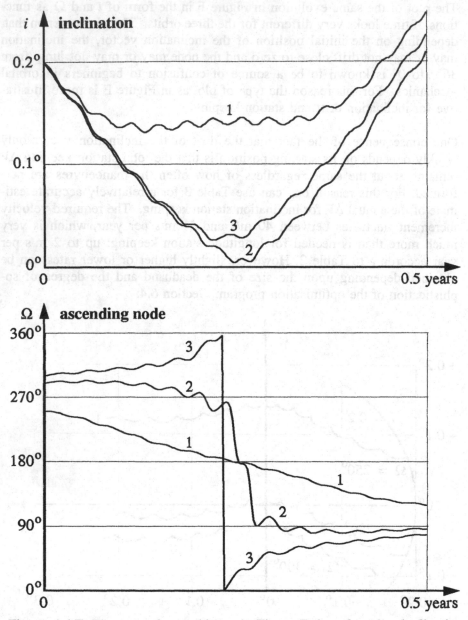

Figure 4.4.F. The same three orbits as in Figure E, but plotted as inclination *i* (above) and ascending node Ω (below) as functions of time during half a year.

For all practical purposes, the inclination vector drift can be considered as independent of the spacecraft longitude and of any longitude manoeuvres that may be executed. This enables the preparation of the long-term inclination manoeuvre strategy to be performed as a stand-alone task, as shown in Section 6.4. For certain missions the strategy can be prepared even before the spacecraft is launched.

4.5 Solar Radiation Pressure

The *solar constant* is the term that refers to the power output of the Sun in the form of electromagnetic radiation. The largest part of the power is radiated as light in the visible and neighbouring wavelengths. The mean radiation intensity near the Earth is about $1.4 \, \text{kW/m}^2$. It changes by only between 0.1% and 0.2% due to variations in the solar activity, whereas the annual change in the Sun-Earth distance contributes to a variation by ±3.3%.

Dividing the power by the velocity of light we obtain the pressure P exerted by the radiation on an orthogonal surface in the neighbourhood of the Earth:

$$P = 4.56 \times 10^{-6} \, \text{N/m}^2$$

This electromagnetic radiation from the Sun should not be confused with the *solar wind*, which consists of electrons and ions, mainly protons, emitted by the Sun and partly trapped by the Earth's magnetic field. The pressure from this particle radiation is many powers of ten smaller than that of the electromagnetic radiation and can be neglected for the purpose of spacecraft motion. The solar panels and other electronic equipment on-board the spacecraft, however, slowly degrade under the influence of the particle radiation.

Changes in solar activity cause a variation in the particle radiation. Of the electromagnetic radiation, mainly the non-visible wavelengths are affected, but they contribute to only a small fraction of the total solar pressure. For this reason one can neglect the solar activity in models of the perturbations in geostationary orbit, but it is important for satellites in low Earth orbits which are exposed to air drag. The density of the upper atmosphere is strongly influenced by variations in the solar activity from the particle, the extreme ultraviolet and the X-ray radiations.

The electromagnetic radiation pressure exerts a force on the spacecraft proportional to its cross-section C:

$$F = PC(1 + \varepsilon)$$

Here ε is the reflectivity coefficient of the surface, lying in the interval $0 < \varepsilon < 1$. The acceleration from the solar radiation pressure

$$dV/dt = PC(1 + \varepsilon)/m$$

depends on the reflectivity and on the cross-section to mass ratio C/m of the spacecraft. This acceleration can become very large for small particles, like debris from exploded satellites. For bodies of the same composition, the mass is roughly proportional to the cube of its linear dimension, whereas the cross-section is proportional to the square, so the acceleration becomes proportional to the inverse of the linear size of the body.

The acceleration is directed away from the Sun and is proportional to the inverse square of the distance, like the gravitational attraction of the Sun. The effect on the motion of the Earth and thereby on MEGSD is, however, quite different. The Earth has a cross-section to mass ratio of 2×10^{-11} m²/kg, whereas typical spacecraft values range from 0.01 m²/kg and upwards, although not higher than 0.1 m²/kg in the foreseeable future. The perturbing acceleration on the spacecraft is then typically of the order of 10^{-7} m/s², whereas the Earth is accelerated by only 10^{-16} m/s².

The result is that, unlike the Sun's gravitational attraction, the total solar radiation acceleration of the order of 10^{-7} m/s² acts on the spacecraft in MEGSD and not only the differential acceleration of Section 4.3. When the spacecraft passes through the shadow of the Earth (Section 5.3) or Moon (Section 5.4), this perturbation is switched off.

Accurate modelling of the solar radiation acceleration is difficult because different surfaces with varying reflectivity and cross-section are often exposed to the sunlight at different parts of the spacecraft orbit. Different types of reflectivity, e.g. specular or diffuse, contribute to the acceleration with components deviating from the direction of the incoming rays. The surface properties in space also deviate from what was measured in laboratory before launch because of outgassing and ageing through radiation damage. However, for most geostationary spacecraft with moderate requirements on orbit accuracy it is sufficient to model the radiation acceleration with one single parameter σ, which we will call *effective cross-section to mass ratio*:

$$\sigma = C(1 + \varepsilon)/m \quad \text{which leads to} \quad dV/dt = P\sigma$$

This parameter will be determined by the orbit determination program at regular intervals during the spacecraft mission for insertion into the perturbation term in the orbit propagation model. The cross-section is defined to be the area of the silhouette of the spacecraft, projected onto a plane perpendicular to the incoming Sun rays, so its size depends on the Sun - spacecraft geometry. Many communications spacecraft have large solar panels that are kept oriented to be perpendicular to the Sun's direction, as projected onto the orbital plane. The size of the cross-section will then vary with cosine of this angle, which usually is close to the declination of the Sun and varies during the year between 0° and 23.4°.

The only orbital elements affected by the solar radiation acceleration are the two components of the eccentricity vector, since the mean effects on the semimajor axis and inclination become zero when averaged over one sidereal day. In Figure A we denote by s_S the sidereal angle of the projection of the Sun's direction onto the orbital plane. When the spacecraft is at the sidereal angle s, the two components in the tangential and radial directions of the velocity differential dV from the the radiation pressure become, respectively:

$$dV_t = \sin(s - s_S)\, dV \quad ; \quad dV_r = -\cos(s - s_S)\, dV$$

From the formulas in Section 3.3 we then obtain the resulting differential equation for the eccentricity vector $\bar{e}$:

$$\frac{d\bar{e}}{dt} = \frac{2}{V}\begin{pmatrix} \cos s \\ \sin s \end{pmatrix}\frac{dV_t}{dt} + \frac{1}{V}\begin{pmatrix} \sin s \\ -\cos s \end{pmatrix}\frac{dV_r}{dt}$$

By integrating this equation over the time interval $\delta t = 2\pi/\psi = 1$ sidereal day, after replacing $dt = ds/\psi$, we obtain the mean drift rate of $\bar{e}$. Since the drift is very small it is not of interest to consider the details of its variation over shorter intervals than a day.

$$\frac{\delta\bar{e}}{\delta t} = \frac{P\sigma}{2\pi V}\int_0^{2\pi}\left[2\begin{pmatrix}\cos s \\ \sin s\end{pmatrix}\sin(s - s_S) - \begin{pmatrix}\sin s \\ -\cos s\end{pmatrix}\cos(s - s_S)\right]ds =$$

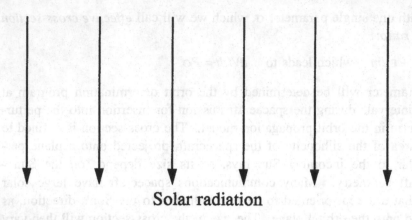

Solar radiation

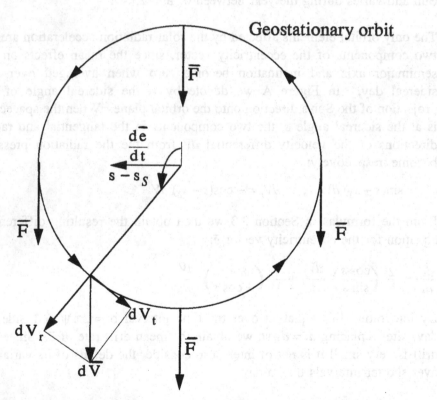

Figure 4.5.A.   Eccentricity vector drift, $d\bar{e}/dt$ due to the solar radiation pressure.

$$= \frac{3P\sigma}{2V} \begin{pmatrix} -\sin s_S \\ \cos s_S \end{pmatrix}$$

The result shows that the mean drift is in the direction $+90°$ from the direction to the Sun. In front of σ there is the factor:

$$3P/2V = 1.9 \times 10^{-4} \, \text{kg/m}^2/\text{day} = 0.58 \times 10^{-2} \, \text{kg/m}^2/\text{month}$$

When σ is greater than the order of 0.01 or 0.02 m²/kg, the resulting $\delta\bar{e}/\delta t$ dominates over the intermediate lunar-gravity perturbation loops, described at the end of Section 4.3. The direction of $\delta\bar{e}/\delta t$ changes throughout the year since the sidereal angle of the Sun changes. If we approximate the Sun's motion as a uniform rotation in the orbital plane, we can solve the resulting difference equation for $\bar{e}$:

$$\bar{e}(t) = \bar{e}_0 + \frac{3P\sigma Y}{4\pi V} \begin{pmatrix} \cos s_S \\ \sin s_S \end{pmatrix}$$

Here $\bar{e}_0$ is the constant of integration and Y is the period of rotation of the Sun $= 1$ year. We have neglected the fact that the angle between the Sun's direction and the orbital plane varies during the year with the declination of the Sun between $0°$ and $23.4°$. The acceleration in the plane should be reduced by a factor cosine of the angle, which means a reduction of up to 8%, at the solstices. Note that this effect is additional to the dependence of the cross-section on the same angle that was mentioned before. The approximate solution shows that $\bar{e}$ moves in a circle with centre $\bar{e}_0$, a period of one year and radius $= 3P\sigma Y/4\pi V$. The coefficient in front of σ is:

$$3PY/4\pi V = 0.011 \, \text{kg/m}^2$$

The vector $\bar{e} - \bar{e}_0$ always points towards the Sun. Figure B shows half a year's free drift of $\bar{e}$, obtained by numerical integration of the orbit. The motion of $\bar{e}$ due to the solar radiation pressure is superimposed on the intermediate-term lunar-gravity perturbation. What is not shown in Figure B is the effect of the longitude station keeping manoeuvre burns. Section 7.3 describes how these burns can be performed to obtain an optimum reduction of the eccentricity.

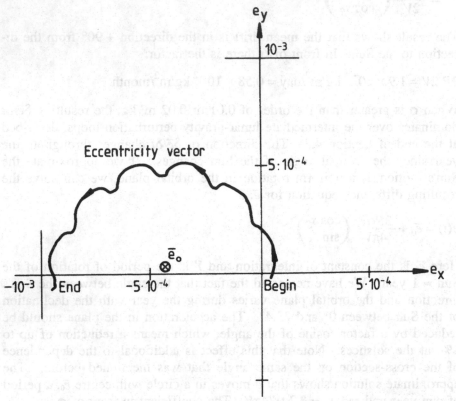

Figure 4.5.B. Eccentricity vector drift during half a year; from spring to autumn. The vector $\bar{e}$ moves along a circle pointing to the Sun with superimposed waves due to the intermediate lunar-gravity perturbation. The cross-section to mass ratio $\sigma = 0.05$ m²/kg.

5. OPERATIONAL PRACTICE

5.1 In-orbit Control

The station keeping criteria for a geostationary spacecraft are normally that the subsatellite longitude and latitude shall be confined inside a rectangular box, Figure A. The sides of this box are called the longitude and latitude *deadbands*, respectively. Since the latitude variations are due to a non-zero orbit inclination one usually expresses the latitude deadband as a maximum allowed inclination i_{max}.

A circular confinement area, as shown in Figure B, may also be prescribed for the longitude and latitude, but this is usually handled like the previous case by using the square box inscribed in the circle. In this way the planning and execution of station keeping manoeuvres can be carried out independently for the longitude and for the inclination, except as regards the in-plane component of inclination thrusts described in Section 6.5.

Additional station keeping criteria may apply to the velocity with which the spacecraft is allowed to move inside the deadband box. These criteria are expressed as upper limits to the orbit eccentricity and to the mean longitude drift rate.

The present and future values of the spacecraft longitude and inclination are only known with a limited accuracy during the operations, since they are calculated by the orbit determination and prediction procedures. Errors in the determined orbit are due to errors in the tracking measurements, as described in Chapter 8. Errors in the orbit prediction are caused by propagation of errors in the determined orbit, combined with the uncertainty in the size of planned orbit manoeuvres, as mentioned in Chapters 6 and 7.

For some missions it is necessary to have a 99% level of confidence in maintaining the station keeping, whereas other missions may be content with a 50% level. The latter case needs no particular attention, whereas the former case requires consideration of a 3σ error margin on the orbit determination and prediction when one plans the station keeping.

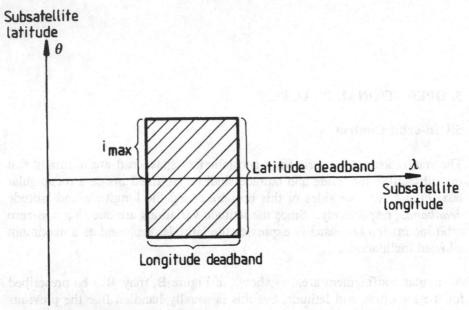

Figure 5.1.A. Station keeping requirement is usually that the subsatellite longitude and latitude shall be confined to a rectangular box.

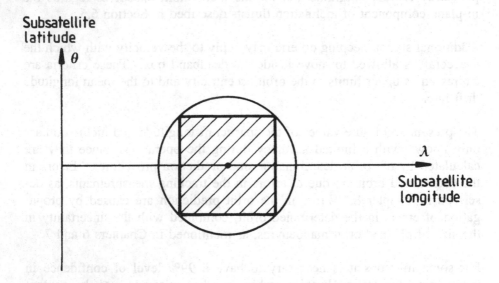

Figure 5.1.B. If the subsatellite point is confined to a circular area one usually uses the inscribed square as station keeping box.

In order to avoid performing stochastic manoeuvre planning, one usually prefers to subtract the error margins, in inclination and longitude, from the deadbands, thereby reducing the size of the station keeping box. An important mission analysis task is to establish what orbit determination accuracy is required for a certain mission. Part of the answer lies in the fact that the remaining box must be wide enough to allow for a reasonably simple station keeping. The complexity of station keeping operations increases drastically with decreasing box sizes.

The spacecraft Operator needs access to predicted orbit information in a convenient form in order to ascertain that the spacecraft is inside the deadband box and to plan station keeping manoeuvres. Section 5.2 gives examples of such printed information. The predicted orbit is also needed for scheduling other operations on the platform and payload, and it is often used to predict the pointing angles of the ground station antennas. Orbital data for the past is sometimes required for a-posteriori processing of payload data or attitude determination.

Station keeping can be formulated as a classical control problem, where the state of the system is expressed by the orbital elements. Measurement of the system is performed by the orbit determination process and the station keeping manoeuvres provide the corrective feedback. From a control theory point of view, however, the system has a very simple on/off control of the same type as a room thermostat. Another simplifying factor is that stochastic control theory can be avoided by the reduction of the station keeping box by the error margins, as described above.

In addition to the routine station keeping manoeuvres, a spacecraft may be required to perform a longitude shift manoeuvre in order, for example, to bring into service an in-orbit spare spacecraft or because of other changes in the mission. Also a longitude station reacquisition manoeuvre may be necessary if the spacecraft has drifted outside the longitude deadband. This could follow upon some mistake in the station keeping performance or as the result of a planned or unplanned attitude change, as in Section 7.5.

As an example of attitude manoeuvres that must be planned with consideration of the influence on the orbit, we mention here the spin axis inversions of ESA's GEOS-2 spacecraft, which was operated between 1978 and 1984. At each equinox the spin axis was changed from north-pointing in summer to south-pointing in winter, and vice versa. The manoeuvres had to be performed with an unbalanced thruster, which perturbed the spacecraft velocity by several m/s.

The most sensitive orbital parameter is the longitude drift rate, so one tries to perform manoeuvres in such a way that the $\Delta \overline{V}$ has no tangential component but only in the other two directions. Another possibility is to split the manoeuvre into two parts with mutually cancelling tangential components. In either case, one needs to determine the orbit soon after and perform a longitude correction thrust, since it is difficult to balance the net tangential component to be zero, in view of manoeuvre errors.

Most three-axis stabilised spacecraft are usually controlled with one axis pointing to the Earth and another axis orthogonal to the orbital plane. Typical requirements on the attitude control of spin-stabilised spacecraft are that the spin axis shall be orthogonal to the plane of, either the orbit, the equator or the ecliptic. In each case, the positive spin axis, as defined below, can be either north pointing or south pointing. Both the spin axis and the orbital plane are drifting slowly, so one may need to perform small attitude station keeping manoeuvres in an analogous way to orbital station keeping.

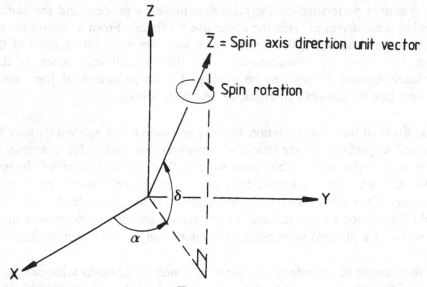

Figure 5.1.C. The unit vector $\overline{Z}$ that is used to indicate the spin axis direction can be expressed in MEGSD by the angles right ascension (α) and declination (δ).

The spin axis direction is represented by a unit vector $\bar{Z}$, aligned with the direction of the angular momentum of the spin. The positive direction of $\bar{Z}$ points like a right-handed screw relative to the spin direction, Figure C. The $\bar{Z}$-vector is usually expressed in the MEGSD co-ordinate system by the two attitude angles *right ascension* (α) and *declination* (δ) in a manner analogous to that in which one expresses the direction of a celestial object from the Earth, Section 2.1. The expression for $\bar{Z}$ in α and δ is, Figure C:

$$\bar{Z} = \begin{pmatrix} \cos\delta \, \cos\alpha \\ \cos\delta \, \sin\alpha \\ \sin\delta \end{pmatrix}$$

Remembering the expression for the three-dimensional orbit inclination vector from Section 2.2, we get the following table for target attitude angles in the three cases:

Spin axis orthogonal to	*North pointing*	*South pointing*
Orbital plane	$\alpha = \Omega - 90°$	$\alpha = \Omega + 90°$
	$\delta = 90° - i$	$\delta = i - 90°$
Equatorial plane	$\alpha =$ undefined	$\alpha =$ undefined
	$\delta = +90°$	$\delta = -90°$
Ecliptic plane	$\alpha = +270°$	$\alpha = +90°$
	$\delta = +66.558°$	$\delta = -66.558°$

A general line of development in space operations is the trend to perform more and more functions on-board in automatic mode, including orbit determination and manoeuvre planning. There is, however, less incentive to perform functions on-board a geostationary spacecraft, which has uninterrupted ground contact, than for other types of missions with only short and infrequent passes over ground stations. When the purpose of the automation is to reduce the manual workload for operations one can equally well implement the automatic system on ground for a geostationary mission. A further reason is that the tracking measurements are produced on ground, although there exist proposals by which the Earth would be tracked by a system on the the spacecraft instead of vice versa, see Section 8.2.

5.2 Computer Program Operations

We will assume here that the spacecraft is controlled from the ground with manual supervision by a spacecraft Operator. In order to perform orbit control, the Operator needs a small or medium-sized computer and a set of orbital control programs. A minimum set consists of programs for, Figure A:

- Tracking data preprocessing
- Orbit determination
- Orbit prediction
- Orbit information
- Longitude station keeping
- Inclination station keeping

The tracking data preprocessing program converts the incoming tracking messages to geometric measurements of distance and direction from one or more ground stations to the spacecraft at discrete times. It also smooths and reduces the raw data, checks the quality and applies calibration and time corrections.

The orbit determination program is normally the largest and most complex of the orbit control programs. It is used to calculate the orbital elements and auxiliary parameters for an epoch by iterated least squares fits of a numerically integrated arc of the orbit to a batch of preprocessed tracking data. This program should be run regularly, at least every week, using eight days of accumulated tracking, including one day of overlap with the previous run. In addition it has to be run about a few days after each manoeuvre in order to evaluate the performance of the thrust and update the orbit prediction. It is useful to run it also just before the planning and execution of longitude manoeuvres, since they depend strongly on the latest value of the longitude drift rate. On the other hand, inclination manoeuvres can be prepared without a very recent orbit determination.

The auxiliary parameters to be determined by the orbit determination program must include at least the solar radiation pressure and the $\Delta \bar{V}$ of the manoeuvre thrusts. The latter can be obtained in principle by determination of the orbit separately before and after the thrust. One then calculates the actual $\Delta \bar{V}$ from the discontinuity in the orbital velocity at the crossing of the two orbits.

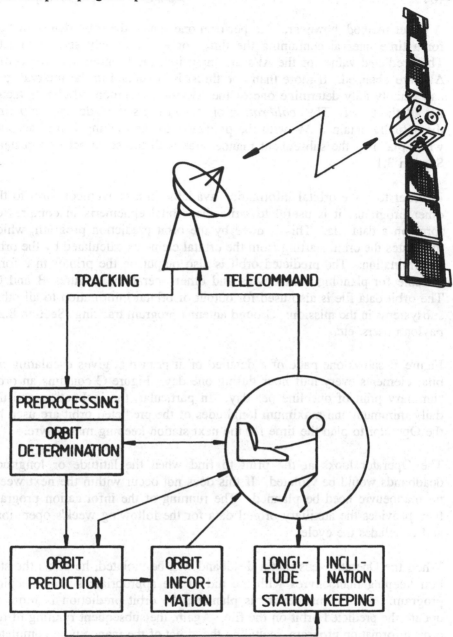

Figure 5.2.A. Schematic illustration of the station keeping control loop.
The Operator runs the six orbit control programs that give the information
needed for station keeping and other control activities.

A better method, however, is to perform one run of the orbit determination for a time interval containing the thrust or a few closely spaced thrusts. The predicted values of the $\Delta \overline{V}$s are input into the program and the actual $\Delta \overline{V}$s are obtained. If more than one thrust is contained in the interval, one can usually only determine one of the $\Delta \overline{V}$s or a common efficiency factor for all the thrusts. The *calibration* of the thrusters is made by comparing the actually obtained $\Delta \overline{V}$ with the predicted $\Delta \overline{V}$ and taking it into account when planning the subsequent manoeuvres with the same set of thrusters, Section 3.1.

In order to make orbital information available in a convenient form to the other programs, it is useful to write the orbital ephemeris in compressed form on a data file. This is done by the orbit prediction program, which integrates the orbit, starting from the orbital elements calculated by the orbit determination. The predicted orbit is also output on the printer in a form suitable for planning manoeuvres and other operations, Figures B and C. The orbit data file is also used for output of orbital information to all other subsystems in the mission: Ground antenna program tracking (Section 8.2), payload users, etc.

Figure B shows one page of a detailed orbit print that gives osculating orbital elements every half hour during one day. Figure C contains an orbit summary print of one line per day. In particular, the inclination and the daily minimum and maximum longitudes of the predicted orbit are used by the Operator to plan the time for the next station keeping manoeuvre.

The Operator looks at the print to find when the latitude or longitude deadbands would be violated. If this does not occur within the next week, no manoeuvre need be planned. The running of the information program then provides the auxiliary orbital data for the following week's operations and concludes the cycle.

When the Operator sees that a deadband will be violated, he plans the station keeping manoeuvres with the aid of the appropriate station keeping program. After a manoeuvre is planned, the orbit prediction is rerun to update the predicted orbit on the file. Again, the subsequent running of the orbit information program, including the effect of the manoeuvre, completes the weekly cycle.

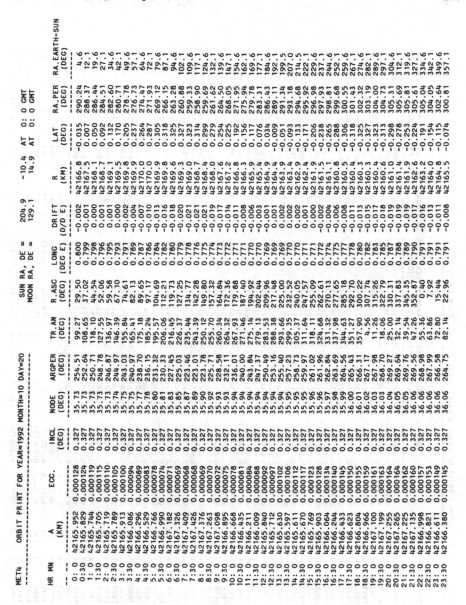

Figure 5.2.B. Typical orbit prediction print, giving osculating orbital elements every half hour, followed by spacecraft right ascension, longitude, longitude drift rate, distance from Earth's centre, latitude, right ascension of perigee and angle between Sun's and Earth's centres, seen from the spacecraft.

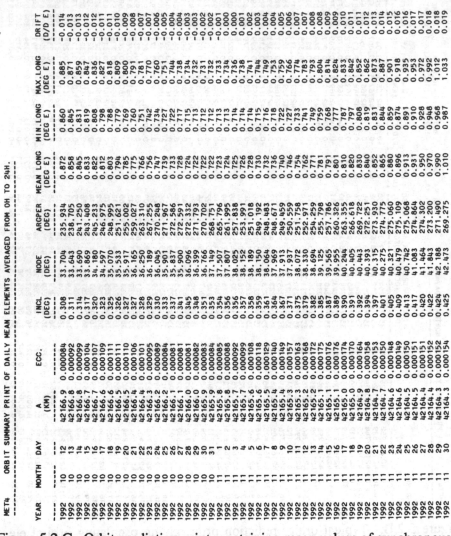

METH   ORBIT SUMMARY PRINT OF DAILY MEAN ELEMENTS AVERAGED FROM 0H TO 24H.

YEAR	MONTH	DAY	A (KM)	ECC.	INCL (DEG)	NODE (DEG)	ARGPER (DEG)	MEAN LONG (DEG E)	MIN.LONG (DEG E)	MAX.LONG (DEG E)	DRIFT (D/D E)	RA.PER (DEG)
1992	10	12	42166.9	0.000084	0.308	33.704	235.934	0.872	0.860	0.885	-0.014	-90.36
1992	10	13	42166.8	0.000092	0.311	33.643	238.705	0.858	0.845	0.871	-0.013	-87.65
1992	10	14	42166.8	0.000099	0.314	33.690	241.250	0.845	0.831	0.859	-0.013	-85.06
1992	10	15	42166.7	0.000104	0.317	33.869	243.408	0.833	0.819	0.847	-0.012	-82.72
1992	10	16	42166.7	0.000107	0.320	34.180	245.233	0.822	0.808	0.836	-0.011	-80.59
1992	10	17	42166.6	0.000109	0.323	34.597	246.975	0.812	0.798	0.827	-0.011	-78.43
1992	10	18	42166.6	0.000111	0.325	35.070	248.995	0.803	0.788	0.818	-0.010	-75.94
1992	10	19	42166.5	0.000110	0.326	35.533	251.621	0.794	0.779	0.809	-0.010	-72.85
1992	10	20	42166.5	0.000101	0.327	35.917	255.002	0.785	0.769	0.800	-0.009	-69.08
1992	10	21	42166.4	0.000106	0.327	36.165	259.027	0.775	0.760	0.791	-0.008	-64.81
1992	10	22	42166.4	0.000095	0.328	36.250	263.310	0.766	0.751	0.781	-0.008	-60.44
1992	10	23	42166.3	0.000089	0.329	36.189	267.255	0.756	0.742	0.770	-0.007	-56.56
1992	10	24	42166.2	0.000084	0.330	36.045	270.045	0.747	0.734	0.760	-0.007	-53.71
1992	10	25	42166.2	0.000081	0.333	35.837	271.967	0.739	0.727	0.751	-0.006	-52.13
1992	10	26	42166.1	0.000081	0.337	35.901	272.586	0.733	0.722	0.744	-0.005	-51.58
1992	10	27	42166.1	0.000082	0.341	35.900	272.591	0.728	0.717	0.738	-0.004	-51.57
1992	10	28	42166.0	0.000083	0.345	36.096	272.332	0.724	0.713	0.734	-0.004	-51.57
1992	10	29	42166.0	0.000084	0.348	36.399	271.793	0.722	0.715	0.732	-0.003	-51.81
1992	10	30	42165.9	0.000085	0.351	36.766	270.702	0.722	0.712	0.731	-0.003	-52.53
1992	10	31	42165.9	0.000088	0.353	37.149	268.751	0.722	0.712	0.732	-0.002	-54.10
1992	11	1	42165.8	0.000092	0.354	37.507	265.796	0.723	0.713	0.733	-0.002	-56.70
1992	11	2	42165.7	0.000099	0.356	37.807	261.995	0.724	0.713	0.734	-0.001	-60.20
1992	11	3	42165.7	0.000108	0.356	38.025	257.838	0.725	0.714	0.736	-0.001	-64.14
1992	11	4	42165.6	0.000118	0.357	38.152	253.991	0.726	0.714	0.738	0.000	-67.86
1992	11	5	42165.6	0.000129	0.358	38.189	251.018	0.728	0.715	0.741	0.000	-70.79
1992	11	6	42165.5	0.000140	0.359	38.150	249.192	0.730	0.716	0.744	0.001	-72.66
1992	11	7	42165.5	0.000149	0.361	38.064	248.483	0.732	0.718	0.749	0.002	-73.45
1992	11	8	42165.4	0.000157	0.364	37.969	248.671	0.736	0.722	0.753	0.003	-73.36
1992	11	9	42165.4	0.000163	0.367	37.913	249.459	0.740	0.727	0.759	0.004	-72.63
1992	11	10	42165.3	0.000168	0.371	37.937	250.559	0.746	0.733	0.766	0.004	-71.50
1992	11	11	42165.3	0.000172	0.375	38.072	251.758	0.754	0.741	0.774	0.005	-70.17
1992	11	12	42165.2	0.000175	0.379	38.330	252.971	0.762	0.749	0.783	0.006	-68.70
1992	11	13	42165.2	0.000176	0.382	38.694	254.259	0.771	0.759	0.793	0.007	-67.05
1992	11	14	42165.1	0.000174	0.385	39.125	255.798	0.781	0.768	0.804	0.007	-65.08
1992	11	15	42165.1	0.000170	0.387	39.565	257.786	0.791	0.777	0.814	0.008	-62.65
1992	11	16	42165.1	0.000164	0.389	39.955	260.326	0.801	0.787	0.824	0.008	-59.72
1992	11	17	42165.0	0.000158	0.390	40.244	263.355	0.810	0.797	0.833	0.009	-56.40
1992	11	18	42165.0	0.000153	0.391	40.405	266.618	0.820	0.808	0.842	0.009	-52.98
1992	11	19	42164.8	0.000150	0.392	40.443	269.722	0.830	0.819	0.852	0.010	-49.83
1992	11	20	42164.7	0.000149	0.394	40.393	272.251	0.840	0.831	0.862	0.011	-47.36
1992	11	21	42164.7	0.000150	0.397	40.315	273.930	0.852	0.844	0.873	0.011	-45.75
1992	11	22	42164.6	0.000148	0.401	40.275	274.775	0.865	0.859	0.887	0.012	-44.95
1992	11	23	42164.6	0.000149	0.405	40.321	275.060	0.880	0.874	0.901	0.013	-44.41
1992	11	24	42164.5	0.000150	0.409	40.479	275.109	0.896	0.891	0.918	0.014	-44.19
1992	11	25	42164.5	0.000151	0.413	40.742	275.068	0.913	0.910	0.935	0.015	-44.05
1992	11	26	42164.4	0.000151	0.417	41.083	274.864	0.931	0.928	0.953	0.016	-44.23
1992	11	27	42164.4	0.000152	0.420	41.464	274.302	0.950	0.948	0.972	0.017	-44.62
1992	11	28	42164.4	0.000152	0.422	41.843	273.200	0.970	0.968	0.992	0.018	-44.96
1992	11	29	42164.3	0.000153	0.424	42.188	271.490	0.990	0.987	1.012	0.018	-46.32
1992	11	30	42164.3	0.000154	0.425	42.473	269.275	1.010	1.007	1.033	0.019	-48.25

Figure 5.2.C. Orbit prediction print, containing mean values of synchronous elements, converted back to classical elements (a, e, i, Ω, ω). Minimum and maximum subsatellite longitude is also given for each day.

SATELLITE : MET4   INFO PRINT FOR DAY NO. 294 = DATE 1992/10/20   (MJD : 15633)   COMPUTED ON 1992/10/20   8:45:42

```
ORBITAL ELEMENTS AT 00:00
SEMI-MAJOR AXIS (KM)      = 42166.0
ECCENTRICITY             = 0.00013
INCLINATION (DEG)        = 0.327
R.A. OF ASC. NODE (DEG)  = 35.73
ARGUMENT OF PERIGEE (DEG)= 254.51
LONGITUDE (DEG E)        = 0.800

MAX. LONGITUDE (DEG E)   = 0.800
MIN. LONGITUDE (DEG E)   = 0.769
LONGITUDE DRIFT (D/D E)  = -0.009
INCLINATION CHANGE (DEG) = 0.000

ATTITUDE AT 00:00
ATTITUDE R.A. (DEG)  = 133.14
SPIN PERIOD (SEC)    = 0.601
ATTITUDE DEC. (DEG)  = 89.80

MANOEUVRE(S)   NONE
```

SOLAR INFORMATION: POSITION AT 00:00   RIGHT ASCENSION = 204.9   DECLINATION = -10.4

ORTHOGONALITY SUN-SATELLITE-EARTH AT 5:42 17:42   SIC ANOMALY AT 13:23:24
RIGHT ASCENSION OF SUN - EARTH, SUN DECLINATION IN ORBITAL SYSTEM, ANGLE SATELLITE - SUN FROM STATION ODN150

HOUR	00:00	01:00	02:00	03:00	04:00	05:00	06:00	07:00	08:00	09:00	10:00	11:00
R. ASC	355.5	340.5	325.5	310.5	295.5	280.5	265.5	250.5	235.5	220.5	205.5	190.5
DECL	-10.4	-10.4	-10.4	-10.5	-10.5	-10.5	-10.5	-10.5	-10.5	-10.5	-10.6	-10.6
ANGLE	162.0	154.4	142.2	128.5	114.2	99.8	85.1	70.4	55.7	41.0	26.2	11.7

HOUR	12:00	13:00	14:00	15:00	16:00	17:00	18:00	19:00	20:00	21:00	22:00	23:00
R. ASC	175.5	160.5	145.5	130.5	115.5	100.5	85.5	70.5	55.5	40.4	25.4	10.4
DECL	-10.6	-10.6	-10.6	-10.6	-10.6	-10.7	-10.7	-10.7	-10.7	-10.7	-10.7	-10.8
ANGLE	4.9	18.7	33.4	48.2	62.9	77.6	92.2	106.8	121.1	135.1	148.2	158.7

LUNAR INFORMATION: POSITION AT 00:00   RIGHT ASCENSION = 129.1   DECLINATION = 14.9

RIGHT ASCENSION OF MOON - EARTH, MOON DECLINATION IN ORBITAL SYSTEM, MOON PHASE

HOUR	00:00	01:00	02:00	03:00	04:00	05:00	06:00	07:00	08:00	09:00	10:00	11:00
R. ASC	286.0	271.8	257.2	242.1	226.6	210.7	194.5	178.1	161.8	145.6	129.8	114.3
DECL	14.2	14.4	14.8	15.0	15.1	15.1	14.9	14.6	14.2	13.7	13.1
PHASE	0.35	0.35	0.35	0.35	0.36	0.37	0.38	0.39	0.40	0.40	0.41

HOUR	12:00	13:00	14:00	15:00	16:00	17:00	18:00	19:00	20:00	21:00	22:00	23:00
R. ASC	99.4	84.8	70.7	57.0	43.5	30.3	17.2	4.2	351.2	338.2	325.1	311.8
DECL	12.5	12.0	11.5	11.0	10.6	10.2	9.9	9.7	9.5	9.4	9.3
PHASE	0.40	0.40	0.39	0.38	0.37	0.35	0.34	0.32	0.30	0.29	0.28	0.26

ECLIPSE INFORMATION: PENUMBRA STARTS   UMBRA STARTS   UMBRA ENDS   PENUMBRA ENDS
NO ECLIPSE TODAY.

TIME OF ASCENDING NODE = 0:25

```
STATION ANGLES AT 00:00
         AZIMUTH   ELEVATION        AZIMUTH   ELEVATION        AZIMUTH   ELEVATION
ODN150   190.67    32.46    KOUROU  93.90     28.58    LANION  174.33    33.89
```

Figure 5.2.D. Orbit information print giving, in particular, the Sun - Moon - Earth - spacecraft geometry and eclipse predictions.

Figure D shows a page of an orbit information program. It gives geometric information about the Sun, Moon and Earth, seen from the spacecraft. It is used by the Operator for setting the parameters of the onboard control loops that control the orientation of three-axis stabilised spacecraft and for scheduling eclipse operations, etc.

The frequency of station keeping manoeuvring depends on the size of the longitude and latitude deadbands. For some missions, only longitude station keeping is needed. For missions where both longitude and inclination station keeping is applied, the longitude thrusts are usually carried out more often than the inclination thrusts. Figure E shows an example of inclination and longitude evolution under the influence of one north and two east thrusts. In this case the in-plane component of the north thrust by chance had the desired influence on the longitude drift change as the additional east thrust that otherwise would have had to be performed.

There exist in principle two different ways to schedule station keeping manoeuvres, both for inclination and longitude:

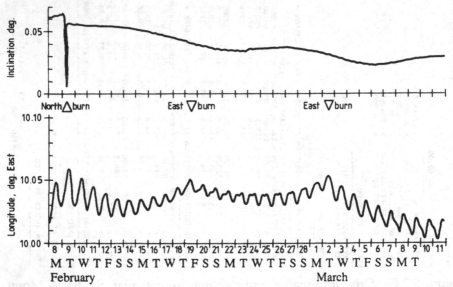

Figure 5.2.E.   Station keeping with one inclination and two longitude manoeuvre thrusts for ESA's Orbital Test Satellite (OTS) during one month in early 1982.

- Wait with the station keeping manoeuvre until the day before the inclination or longitude exceeds the deadband. This provides the longest time between manoeuvres.

- Perform the station keeping manoeuvre according to a regular schedule with a cycle of one or two weeks. This simplifies the planning of working hours for the Operator.

A typical routine station keeping schedule for ESA's ECS satellites contained two longitude manoeuvres and one inclination manoeuvre in a fixed 4 week cycle, as follows. Tracking data was collected continuously, also during the weekend.

Week 1, Monday:	Orbit determination and prediction.
	Check deadbands for next 2 weeks.
Week 1, Friday:	Orbit determination and prediction.
	Prepare an inclination manoeuvre for Monday.
Week 2, Monday:	Execute the inclination manoeuvre.
Week 2, Tuesday:	Quick orbit determination and drift check.
Week 2, Wednesday:	Orbit determination and prediction.
	Prepare a longitude manoeuvre for next day.
Week 2, Thursday:	Execute the longitude manoeuvre.
Week 2, Friday:	Quick orbit determination and drift check.
Week 3, Monday:	Orbit determination and prediction.
	Check deadbands for next 2 weeks.
Week 4, Monday:	Orbit determination and prediction.
	Prepare a longitude manoeuvre for next day.
Week 4, Tuesday:	Execute the longitude manoeuvre.
Week 4, Friday:	Orbit determination and prediction.
	Check deadbands for next 2 weeks.

When several spacecraft of the same series are operated by the same centre, the station keeping cycles may be interleaved. Some missions have a fixed cycle for the longitude and a flexible schedule for the inclination station keeping. It is convenient always to make an inclination manoeuvre a few days before a planned longitude station keeping in order to include in the latter a compensation for the inevitable along-track errors, Section 6.5. In the days between the manoeuvres, new tracking data is collected, the new orbit is determined and the longitude manoeuvre is prepared. Very short cycles caused by tight deadbands are considerably more complicated and have to be worked out on a case-by-case basis.

5.3 Eclipse by Earth

Twice per year, at spring and autumn, the Sun moves through the equatorial plane, Figure A. Since the orbital plane is kept reasonably close to the equatorial plane this implies that the spacecraft passes through the Earth's shadow once per day during this period. This is called an eclipse of the Sun by the Earth, but it is really the same as the night, as experienced by an observer on the surface of the Earth.

Eclipses are a nuisance to spacecraft missions for several reasons: the on-board temperature changes, the solar cells cease to give power and the Sun reference direction is lost. Depending on the spacecraft's design, various devices on board have to be switched during the eclipse, with a safety margin before and after: heaters are switched on, parts of the payload are switched off and battery power is switched on. Often manoeuvres are prohibited in this time interval.

For the purpose of eclipse calculations one can consider the Earth to be spherical with a radius $R = 6378.144$ km, corresponding to an angle of $8.70°$ as seen from a geostationary distance. The Sun's disc is seen with the diameter at the angle of $0.5°$, so there is a penumbra of $0.5°$ at the edge of the dark region, the umbra, Figure B. The spacecraft moves $1°$ in 4 minutes, so the maximum eclipse duration is 71.5 minutes, of which 2 minutes at the beginning and 2 at the end are in the penumbra. The Sun's declination changes by about $0.4°$ per day at the eclipse season, so there is one pure penumbra passage, of maximum duration 23.5 minutes, that starts and ends each season.

In this context, it is convenient to introduce the concept of *local spacecraft time*. This is the same as the local solar time on Earth at the longitude that coincides with the subsatellite longitude. The eclipses always occur symmetrically around the local spacecraft midnight. When choosing the operational longitude for a communications spacecraft, in particular when it is used for television broadcasting, one must keep the eclipse hours in mind. By selecting the spacecraft longitude further west than the location of the payload users (= TV viewers), one can avoid eclipses and payload switch-off during evening broadcasting.

Figure C shows the dates and duration of the shadow passes of a spacecraft with orbit inclination zero. The corresponding dates are almost the same for every year with a maximum of one day difference. Each eclipse season lasts for 46 days, from February 26 to April 13 and from August 31 to October 16. When the inclination is not zero, both the time of the eclipse season and its total length change, although the duration of the longest eclipse is the same. Figure D shows the shifts of the mid date of the eclipse season, which coincides with the date of the longest eclipse. The duration of the eclipse season is plotted in Figure E. The season is longer, the smaller the angle is between the orbital plane and the ecliptic plane.

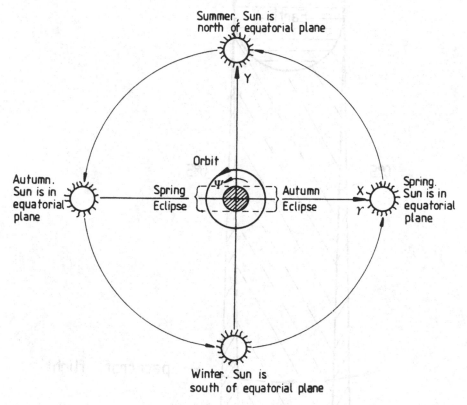

Figure 5.3.A. Eclipses of the Sun by the Earth are experienced by a geostationary spacecraft around the spring and autumn equinoxes.

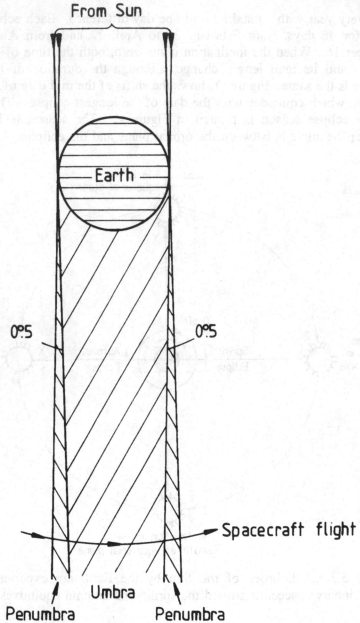

Figure 5.3.B. At the time of maximum eclipse the spacecraft passes through a shadow almost two earth radii long, the umbra. The penumbra is the lighter shadow at the edges caused by the Sun's view angle of 0.5° (not drawn to scale).

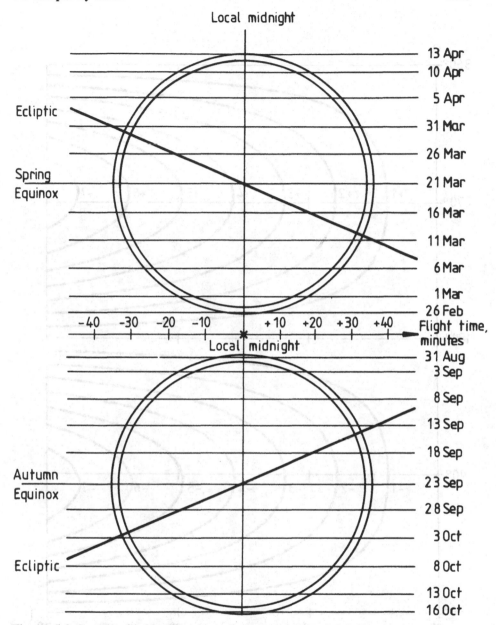

Figure 5.3.C. The flight of a geostationary spacecraft with $i = 0$ through the penumbra (outer circle) and umbra (inner circle) on different days near the spring and autumn equinoxes. Flight is from left to right and the scale shows the flight time in minutes from local midnight.

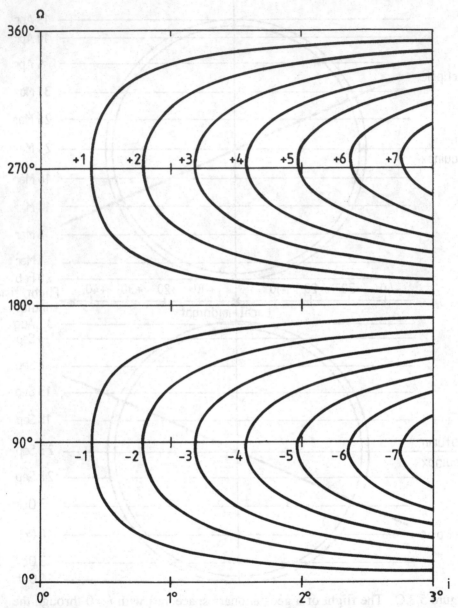

Figure 5.3.D. The level curves show by how many days, counted respectively from March 21 and September 23, the mid date of the eclipse season is shifted, as functions of i and Ω of the spacecraft orbit.

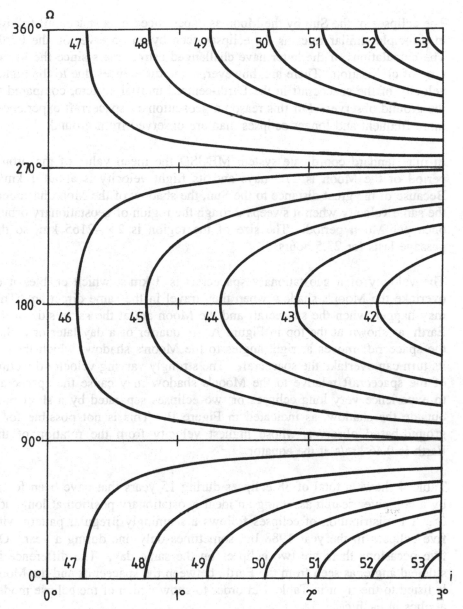

Figure 5.3.E. The level curves show the length, in days, of the eclipse season as a function of i and Ω of the spacecraft orbit.

5.4 Eclipse by Moon

The eclipses of the Sun by the Moon as experienced by a spacecraft follows
in principle similar rules as the eclipses seen by an observer on the Earth.
The calculations of the latter have challenged astronomers since the beginning of civilisation. There are, however, some differences due to the higher
velocity of the spacecraft in the Earth-centred inertial system, compared to
the ground observer. For this reason a geostationary spacecraft experiences
more frequent and longer eclipses than are observed from ground.

In our standard coordinate system MEGSD the mean value of the orbital
period of the Moon is 27.3 days and its flight velocity is about 1 km/s.
Because of the great distance to the Sun, the shadow of the Moon has about
the same velocity when it sweeps through the region of geostationary orbits,
once per Moon period. The size of the region is 2×42165 km, so the
passage lasts for 23.5 hours.

The velocity of a geostationary spacecraft is 3 km/s, which enables it to
overtake the Moon's shadow when they travel in the same direction. This
may happen when the spacecraft and the Moon are at the same side of the
Earth, as shown at the top in Figure A. A quarter of a day later or earlier
the spacecraft moves at right angles to the Moon's shadow, which then in
its turn can overtake the spacecraft. The strongly varying velocity direction
of the spacecraft relative to the Moon's shadow may cause the spacecraft
to experience very long eclipses or two eclipses separated by a short time
outside the shadow, as indicated in Figure B. This is not possible for a
ground-based observer, whose highest velocity from the rotation of the
Earth is 0.46 km/s at the equator.

Table 4 shows a total of 38 eclipses during 15 years that have been found
by a computer search assuming an ideal geostationary position at longitude
0°. The distribution of eclipses follows a seemingly irregular pattern with
five eclipses in the year 1988 but sometimes only one during a year. On
five occasions there are two eclipses on the same day. The difference in
sidereal angle, as seen from the Earth, between the spacecraft and the Moon
is listed to the right in Table 4 in order to show which of the eclipse modes
applies in each case.

The description of the eclipses so far are valid both for umbra and
penumbra. Details of the two is drawn in Figure C, which schematically

shows the geometry of the shadow cast by the Moon. A full eclipse is experienced by the spacecraft only inside the umbra, which has the shape of a long narrow cone. The tip of the cone is located in the region through which the spacecraft passes, and it may fly on either side of it. Around the umbra lies the wider cone of the penumbra, the cross-section of which has a radius of about 3500 km in the applicable region.

Below is shown an overview of the relevant geometric dimensions. The distance of the Earth from the Sun and of the Moon from the Earth vary because of the orbital eccentricities. On the other hand, because of the great distance to the Sun one can consider that it is the same from the Earth, from the Moon and from the spacecraft.

The geometric dimensions for the Sun, Moon and spacecraft

Distance Sun - Earth (or Moon)	= from 147×10^6 to 152×10^6 km
Distance Moon - centre of Earth	= from 356000 to 414000 km
Distance Moon - spacecraft	= from 314000 to 456000 km
Radius of Sun	= 696000 km
Radius of Moon	= 1738 km
Distance Moon - tip of umbra cone	= from 367000 to 381000 km
Visible radius of Sun from spacecraft	= from 0.262° to 0.271°
Visible radius of Moon from spacecraft	= from 0.218° to 0.317°

The list above shows that, as seen from the spacecraft, the Sun may appear either larger than or smaller than the Moon (Figure D), depending on which side of the tip of the umbra cone the spacecraft flies, Figure C. The calculated radii of the discs is listed in Table 4. A full umbra, when the Sun vanishes completely behind the Moon's disc (Figure D bottom), is very unusual and does not appear at all in Table 4. The nearest full eclipse is the second eclipse on 1990/08/20, where only 0.01° of the Sun's disc is not occulted. Separate calculations show that the full eclipse would have been seen from a satellite at longitude 1° at the same occasion.

On 1988/09/11 the Sun, Moon and spacecraft were almost completely aligned. However, the disc of the Moon was seen completely inside but smaller than the Sun's disc, Figure D centre. This is a very rare type of penumbra. Most eclipses consist of a penumbra where a part of the Moon is seen to pass in front of the Sun, Figure D top.

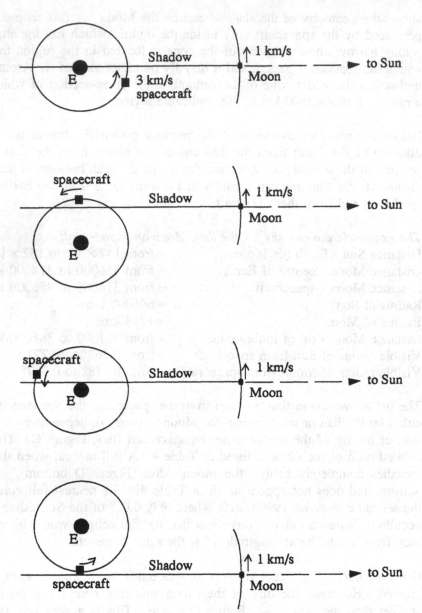

Figure 5.4.A. Four different geometries just before the start of an eclipse by the Moon. The sidereal angle of the spacecraft, relative to the Moon, is 0°, 90°, 180° and 270°, respectively.

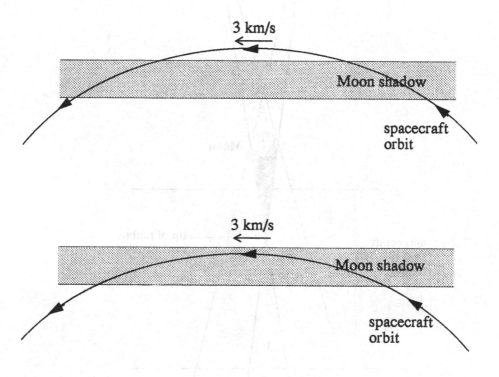

Figure 5.4.B. Two closely spaced eclipses (top) or one long eclipse (bottom) may occur when the spacecraft flight direction is approximately aligned with the shadow. The motion of the shadow can be either up or down in the drawings.

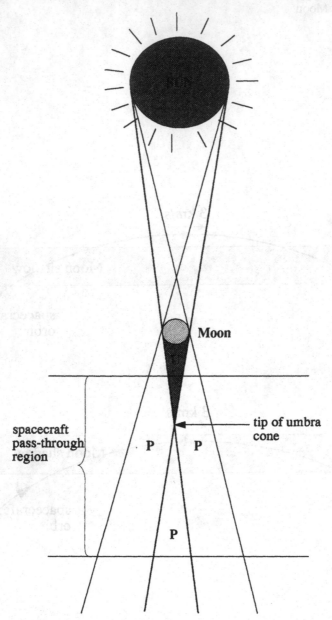

Figure 5.4.C. Schematic overview, not to scale, of the umbra (U) and penumbra (P) shadows cast by the Moon.

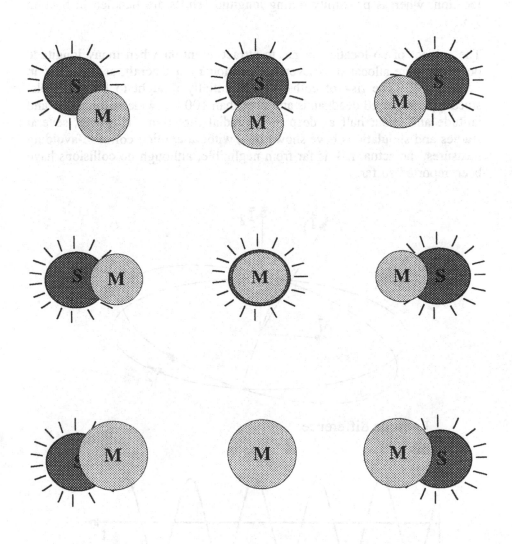

Figure 5.4.D. Top row: the most common type of penumbra; centre row:
a very unusual type of penumbra; bottom row: a complete umbra. To the
right and left are shown the geometries before and after the maximum
occultation in the centre column.

5.5 Co-location

This and the next section deal with methods for station keeping during co-location, whereas proximity during longitude shifts are handled in Section 7.5.

The subject of co-location began to attract attention when many longitude positions were allocated to several geostationary spacecraft, as described in Section 1.3. The risk of collision was initially thought to be negligible, since typical shared deadbands are more than 100 km wide in longitude and latitude and about half as deep in the radial direction. However, several studies and simulations have shown that, without explicit collision-avoiding measures, the actual risk is far from negligible, although no collisions have been reported so far.

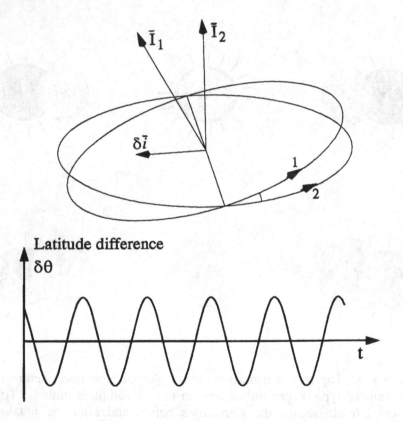

Figure 5.5.A. Two orbits with different inclinations intersect at two points.

The reason is that co-located spacecraft are normally not uniformly distributed within the deadband. Instead, they often fly in closely similar orbits and are operated according to the same optimal station keeping strategy, in particular when they are of the same design. The exact risk of collision is difficult to quantify, and various studies and simulations performed produce highly varying results depending on the assumptions made. Below is shown the result of a very simple calculation of the collision risk for two spacecraft with different inclinations but where no explicit separation strategy is applied. The risk occurs twice per sidereal day when both spacecraft pass the crossing of the two orbital planes, Figure A.

One can guess the collision risk at each plane crossing fly-by in a highly simplified way by comparing the separation, in two directions at the crossing, with the spacecraft size. The latter is typically up to a few tens of metres, whereas the former could be of the order of kilometres in the radial and along-track directions if they are operated according to the same longitude station keeping strategy. The estimated collision probability then lies between 10^{-4} (pessimistic case) and 10^{-5} (optimistic case).

The accumulated risk is shown in the table below. The two columns to the right list the probability of having at least one collision during the mission, as a function of the mission duration.

days =	years	pessim.	optim.
0.5 =	0.00	0.0001	0.00001
1 =	0.00	0.0002	0.00002
2 =	0.01	0.0004	0.00004
4 =	0.01	0.0008	0.0001
8 =	0.02	0.0016	0.0002
16 =	0.04	0.0032	0.0003
32 =	0.09	0.0064	0.0006
64 =	0.18	0.0127	0.0013
128 =	0.35	0.0253	0.0026
256 =	0.70	0.0499	0.0051
512 =	1.40	0.0973	0.0102
1024 =	2.80	0.1852	0.0203
2048 =	5.61	0.3361	0.0401
4096 =	11.21	0.5592	0.0787

Collision risk as a function of mission time.

The result is a quantitative confirmation of the old saying that "also the unlikely event is, with a high probability, bound to happen sooner or later". It is obvious from the table that only the optimistic case provides an acceptably low collision risk for long missions, which often may last more than 10 years.

Most vulnerable to collisions are large communications spacecraft with protruding solar panels in the north-south direction. Because of the low stiffness of the constructions, even an impact velocity as low as 1 m/s would cause considerable damage. The most likely contact point is the tip of a solar panel, causing one panel on each satellite to break or bend. In either case, the result would lead to loss of power and the impact would upset the attitude stability. The latter can be regained, but the high torque caused by the differential solar radiation pressure from one damaged panel would probably make it impossible to maintain the attitude balance in the long run.

As a result of the new awareness of the risk, elaborate station keeping methods have been studied in order to minimise the potential hazard of mutual collision. However, the methods have often been difficult to carry out in practice, in particular when different control centres are involved. An additional problem is often the limited accuracy of the tracking systems, which may originally have been designed only for single station keeping inside the whole deadband. It is then discovered too late that the co-location requires tighter orbit control for each participating spacecraft than foreseen in the mission design.

There are several different ways to perform co-location, and the selection of the method should be based upon the particular characteristics of the participating spacecraft and their ground segments. For some methods it is preferred that all spacecraft are of the same design, whereas others can be used if they are different. The following properties should be considered:

• Number of participating spacecraft.

• Number of control centres.

• Size of deadband, in longitude and latitude.

• Duration of station keeping cycle, in longitude and inclination.

• Orbit determination accuracy.

- Magnitude of solar radiation pressure.

- Thruster accuracy.

- Constraints on thruster firings, e.g. maximum duration, maximum number of cold starts, Sun - Earth angle, etc.

Among different ways to deal with co-location one can distinguish mainly four different approaches:

1. No collision avoiding measures.
The most common approach so far has been to ignore the collision risk completely and operate each co-located spacecraft independently. One can assume that this mode is applied in particular when they are handled by different control centres. It may work for a short time and as long as the co-located members are relatively few. This is probably the reason why there are no collisions reported yet.

2. Uncoordinated co-location with collision checking.
In this mode, each co-located spacecraft is operated to a certain degree independently from the other members, except that on a regular basis and before and after each orbit manoeuvre the predicted distances to the others are checked. A minimum safe distance must have been agreed upon beforehand, the size of which depends on the expected orbit accuracy. If there is an unacceptable proximity predicted, the spacecraft Operator has to modify a planned manoeuvre or perform an extra avoidance manoeuvre. In either case, one can only perform manoeuvres on one spacecraft at a time. The Operator shall then determine the new orbit and check that it is safe before any new manoeuvre is made for any one of the spacecraft. The minimum time between manoeuvres then becomes about 1.5 to 2 days. Frequent information exchange about orbit and manoeuvre planning is necessary, which makes the operations cumbersome when different control centres are involved.

3. Co-location by separation.
Here each spacecraft has its own dedicated region permanently allocated, inside which its station keeping is performed. The next sections shows how different orbital elements can be used for separation. It may be a subset of the total station keeping deadband or an area in the plane of the eccentricity vector, the latter possibly combined with a separation in the plane of the inclination vector. In theory, each spacecraft Operator could perform

his station keeping without exchange of information with the other co-located members if he could rely upon each one to keep to his allocated region. A sufficient safety margin must surround the regions in order to absorb the expected errors of the determined orbit and the predicted manoeuvres. Each dedicated region must accommodate both the short-term orbital librations and the drift during a station keeping cycle, which makes this method impractical when the deadbands are narrow.

4. Coordinated station keeping.

With a narrow deadband one can, instead of 3 above, perform the station keeping with different targets for the different spacecraft according to a pre-defined schedule. As before, the target off-set can be in the longitude or in the plane of the eccentricity vector, optionally combined with a separation in the plane of the inclination vector. The advantage, compared to method 3 above, is that one can make use of the fact that all the members have the same short-term orbital librations. When they are in the same phase of the station keeping cycle the nominal drift is also the same. This motion does not influence the inter-spacecraft distance and can now be excluded from the separation margin. During the station keeping cycle, the members must exchange orbital information and perform proximity checks, but a proximity avoidance manoeuvre is only needed in case of a dangerous approach. Normal manoeuvre errors are otherwise corrected for only at the subsequent station keeping manoeuvre. A practical way to implement this method is to declare one of the members to be the "master", which performs its station keeping independently and relative to which the other co-located members shall manoeuvre.

Whichever approach to co-location operations is selected, the complications increase rapidly with the number of spacecraft and with the number of control centres involved. The greatest proximity risk occurs briefly after an orbit manoeuvre due to the intrinsic inaccuracy in the propulsion system. The critical period lasts until new tracking data has been collected and the new orbit is determined and predicted. In method 2 one must also check the future proximity to all other co-located spacecraft, whereas in method 3 it is sufficient to check that it will stay inside its allocated region. If necessary, one Operator must then plan and execute a correction manoeuvre. This whole process lasts about 2 or 3 days with a normal ground tracking system. Section 5.7 shows how one can take the thruster dispersion into consideration in the safe planning of a nominal manoeuvre and also how to prepare a proximity avoidance manoeuvre.

5.6 Separation Methods

The mathematical calculations in this section mostly apply to two co-located spacecraft. The extension to a larger number n can be performed in principle by repeating the same type of calculation $0.5\,n(n-1)$ times, once for each pair of spacecraft.

In order to describe the motion of two closely co-located spacecraft relative to one another we will use here the linearised representation introduced in Section 2.3. This is a sufficiently good approximation for showing in a qualitative way which properties of the orbits can be used for avoiding a too close proximity. The omission of the natural perturbations in the equations of relative motion can be justified by the fact that almost the same perturbations act on both spacecraft. Only if the cross-section to mass ratio is different will there be a slowly progressing differential perturbation on the eccentricities of the two orbits.

The synchronous orbital elements for the two spacecraft, 1 and 2, are here denoted by, respectively: $(\lambda_{01}, D_1, \bar{e}_1, \bar{i}_1)$ and $(\lambda_{02}, D_2, \bar{e}_2, \bar{i}_2)$ and the difference between the elements by:

$$\delta\lambda_0 = \lambda_{01} - \lambda_{02}; \quad \delta D = D_1 - D_2; \quad \delta\bar{e} = \bar{e}_1 - \bar{e}_2; \quad \delta\bar{i} = \bar{i}_1 - \bar{i}_2$$

The difference between the spacecraft positions is then obtained by subtracting one linearised unperturbed equation of motion from the other. Because of the linearity, the form of the relative motion, of spacecraft 1 relative to spacecraft 2, is of the same form as for a single spacecraft of Section 2.3. On the right hand side below, one can take s to be the sidereal angle that corresponds to the mid-point of the shared longitude deadband for the two co-located members.

$$\delta r = r_1 - r_2 = -A(2/3\,\delta D + \delta e_x\,\cos s + \delta e_y\,\sin s)$$

$$A\delta\lambda = A(\lambda_1 - \lambda_2) = A\delta\lambda_0 + A\delta D(s - s_0) + 2A(\delta e_x\,\sin s - \delta e_y\,\cos s)$$

$$A\delta\theta = A(\theta_1 - \theta_2) = -A(\delta i_x\,\cos s + \delta i_y\,\sin s)$$

The instantaneous inter-satellite distance

$$d = \sqrt{\delta r^2 + (A\delta\lambda)^2 + (A\delta\theta)^2}$$

contains one contribution each from the separation in, respectively, the radial, the along-track and the out-of-plane direction. Some orbital elements must be different for the two orbits in order to ensure spacecraft separation. The inclination vector alone is not sufficient, since the orbital planes always cross along a line that is orthogonal to the differential two-dimensional inclination vector $\delta \bar{i}$, as shown in Figure 5.5.A.

One can now identify the following six basic modes of separating two, or more, co-located spacecraft within a common allocated station keeping deadband.

Mode 1. Complete longitude separation.
This is the simple method of splitting the longitude deadband into smaller deadbands, of which one is allocated to each participating spacecraft. It is possible only if the size of the combined deadband is wide and the number of spacecraft is small. This mode is not really a co-location, since each spacecraft is station kept inside its own partition, independently of the others. In terms of orbital elements, it implies a large value of $\delta \lambda_0$ and a small eccentricity and longitude drift rate for each spacecraft during the station keeping cycle according to:

$$\delta \lambda_0 > 2(e_1 + e_2) + (|D_1| + |D_2|) \left[\max_s |(s - s_0)| \right]$$

Mode 2. Longitude separation during drift cycle.
This mode also splits the longitude deadband, but into partly overlapping regions, which are occupied by different members during different parts of the station keeping cycle as shown in Figure A. It is an example of coordinated station keeping according to method 4 of the preceding section. Longitude manoeuvres shall be made on the same day for all spacecraft so that the longitudes propagate during the cycle according to essentially the same parabola, although with a constant longitude off-set. The eccentricities do not need to be correlated as long as they are sufficiently small so as not cause too large librations in the longitude, but it is also possible to combine this mode with mode 3 below. It is best suited for spacecraft with low solar radiation pressure. In terms of orbital elements, one targets for a given longitude separation $\delta \lambda_0$ and the same drift rate:

$$\delta \lambda_0 > 2(e_1 + e_2) \quad ; \quad \delta D = 0$$

Mode 3. Longitude separation during eccentricity libration.
Here the longitude deadband is also split into partly overlapping regions, but they are occupied by the co-located members during different times of the sidereal day. This is caused by the longitude librations from the eccentricity, Figures B and C, so the mode is suitable for spacecraft with high, but nearly equal, solar radiation pressures. All the members must be in the same phase in their longitude librations, which is in agreement with the requirement of the Sun-pointing-perigee station keeping strategy of Section 7.3. Manoeuvres for the co-located spacecraft should be made at approximately the same time, but some deviation must be allowed in order to correct for errors from the preceding station keeping cycle manoeuvres.

It is possible to permit relatively large values of $\bar{e}_1 \approx \bar{e}_2$, but the station keeping shall target for the following separation of the orbital elements:

$$\delta\lambda_0 > 0 \quad ; \quad \delta D = 0 \quad ; \quad \delta\bar{e} = 0$$

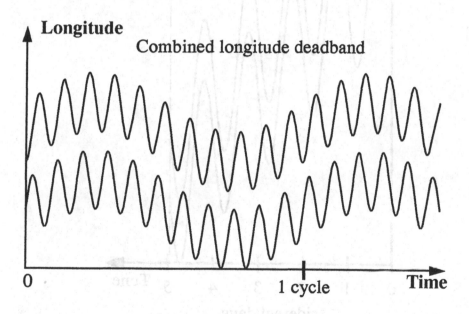

Figure 5.6.A. Correlated station keeping for two spacecraft according to mode 2 with partly overlapping longitude deadbands during the longitude cycle. The plot shows the two subsatellite longitudes versus time during 1.5 cycles.

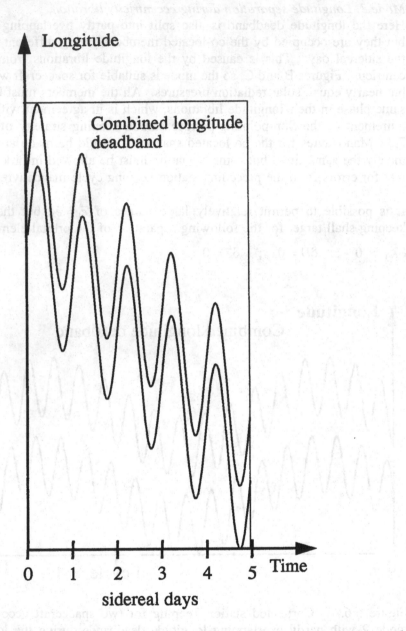

Figure 5.6.B. Co-location of two spacecraft according to mode 3 with partly overlapping deadbands for the longitude librations from the eccentricity. The plot shows the two subsatellite longitudes versus time during 5 days.

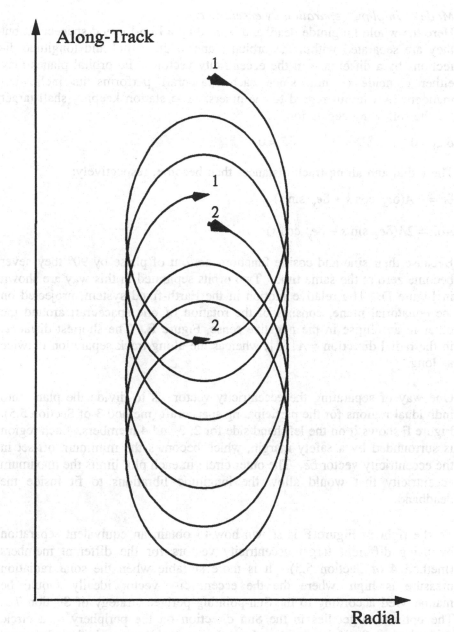

Figure 5.6.C. Co-location of two spacecraft according to mode 3 for 3 days, showing the spacecraft motion projected on the equatorial plane.

Mode 4. In-plane separation by eccentricity.
Here the whole longitude deadband is used by all co-located spacecraft, but
they are separated within the orbital plane in the radial and longitude directions by a difference in the eccentricity vector. The orbital planes may
either coincide or not, since each spacecraft performs its inclination
manoeuvres without regard to the others. The station keeping shall target
for the following separation:

$$\delta \lambda_0 = 0 \quad ; \quad \delta D = 0 \quad ; \quad \delta \bar{e} \neq 0$$

The radial and along-track distances then become, respectively:

$$\delta r = - A(\delta e_x \cos s + \delta e_y \sin s)$$

$$A \delta \lambda = 2A(\delta e_x \sin s - \delta e_y \cos s)$$

Because their sine and cosine functions are out of phase by 90° they never
become zero at the same time. Two orbits separated in this way are shown
in Figure D. The relative motion in the Earth-fixed system, projected on
the equatorial plane, consists of the rotation of one spacecraft around the
other in an ellipse in the negative sense, Figure E. The shortest distance,
in the radial direction $= A |\delta \bar{e}|$, whereas the along-track separation is twice
as long.

One way of separating the eccentricity vectors is to divide the plane into
individual regions for the participating spacecraft (method 3 of Section 5.5).
Figure F shows it on the left hand side for 2, 3 and 4 members. Each region
is surrounded by a safety margin, which becomes the minimum off-set in
the eccentricity vector $\delta \bar{e}$. The outer circle in each plot limits the maximum
eccentricity that would allow the longitude librations to fit inside the
deadband.

To the right in Figure F is shown how to obtain an equivalent separation
by using different target eccentricity vectors for the different members
(method 4 of Section 5.5). It is more suitable when the solar radiation
pressure is high, where the the eccentricity vector ideally should be
manoeuvred according to the Sun-pointing-perigee strategy of Section 7.3.
The optimal target lies in the Sun direction on the periphery of a circle
centred at $\bar{e} = 0$. If the members have the same solar perturbation, the target
circles will get the same radius. By off-setting the centres from 0 in different directions one can ensure the corresponding off-set of each individual
eccentricity vector target. The size of the target circles may now have to

be smaller than the optimal value for a single spacecraft in order to fit them all inside the circle of maximum eccentricity. The price to be paid for co-location is then the extra fuel for multiple-thrust longitude manoeuvres to reduce the eccentricity.

There exist also many other ways to separate the eccentricity vectors in the plane. In the examples of Figure F, one could equally well rotate all vector off-sets $\delta \bar{e}$ by the same angle in the plane. The general principle for arranging the eccentricity vectors is that the configuration shall be as compact as possible in order to fit inside the circle of the maximum allowed eccentricity but at the same time maximise the smallest vector difference $\delta \bar{e}$ between all pairs of the participating spacecraft.

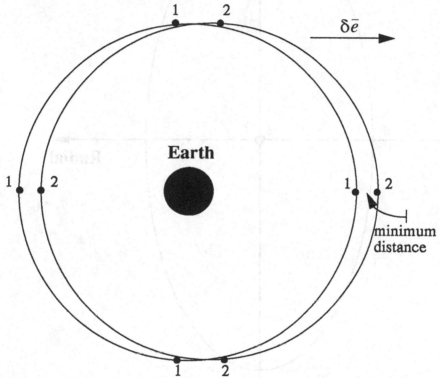

Figure 5.6.D. Equator plane separation by eccentricity. The orbital plane is seen from north in the inertial system with the position of spacecraft 1 relative to spacecraft 2 aligned opposite to the differential eccentricity vector $\delta \bar{e} = \bar{e}_1 - \bar{e}_2$.

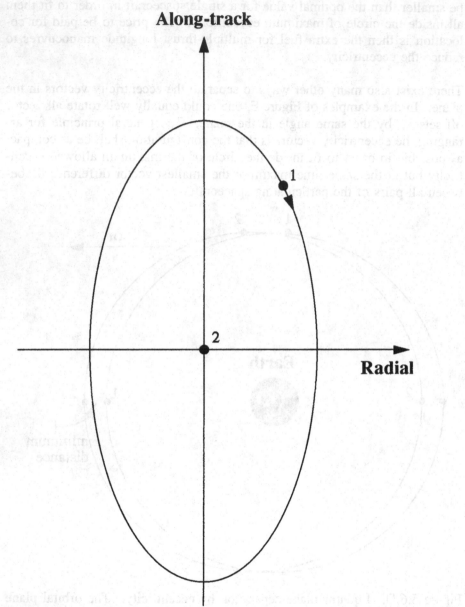

Figure 5.6.E. The same separation as in Figure D, but showing the motion of spacecraft 1 relative to spacecraft 2 in the Earth-rotating system. The motion, always in the negative sense, is along an ellipse that is twice as long, in the along-track direction, as wide.

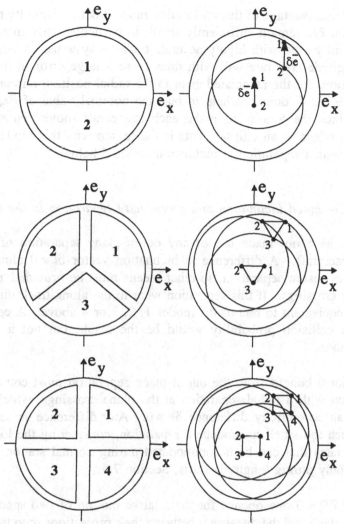

Figure 5.6.F. Co-location of 2, 3 and 4 spacecraft with in-plane separation by the eccentricity, mode 4. Left is shown the plane of eccentricity vectors divided into dedicated regions for separation according to method 3 of Section 5.5. Right is shown the same result by method 4 of Section 5.5. Each spacecraft follows the Sun-pointing perigee strategy but with the target circle off-set from the centre. The 2, 3 and 4 circle centres are marked, as well as points on the periphery in the direction 45°, corresponding to a time in in early May. The outer circle in each plot limits the maximum allowed eccentricity.

The major disadvantage of this co-location mode 4 is the difficulty to ensure that δD and $\delta\lambda_0$ remain sufficiently small, in particular after manoeuvres, so it is mainly used with highly accurate tracking systems. A small error in the longitude drift rate will with time cause a large error in the along-track position, but the associated error in the radial position remains small, in comparison. It does not help to bias the nominal value of $\delta\lambda_0$ by any amount, since the aim is to make each spacecraft move symmetrically around the others in an ellipse. That is the reason why this mode is often combined with a separation in inclination; mode 5 below.

Mode 5. Combined inclination and eccentricity separation in the meridian plane.

So far we have not made use of any out-of-plane separation of the co-located spacecraft. A difference in inclination vector $\delta\bar{i} \neq 0$ alone is not sufficient to ensure separation, so other means must be provided at the orbital plane crossings. If this separation were in the along-track direction it would be equivalent to one of the modes 1, 2, 3 or 4 above. A certain reduction in collision probability would be the result, but not a new co-location mode.

The additional benefit from the out-of-plane separation must come from a combination with a radial separation at the plane crossings, which is provided by an eccentricity difference $\delta\bar{e} \neq 0$. Any difference in semimajor axis is much too small to produce a separation, considering the low longitude drift rates that can be accommodated during normal station keeping. It is used only during longitude shifts, Section 7.5.

By setting $\delta D = 0$ one obtains for the relative motion of two spacecraft in the $(r, A\theta)$-plane and the distance d between their projections onto this plane:

$$\delta r = - A(\delta e_x \cos s + \delta e_y \sin s)$$

$$A\delta\theta = - A(\delta i_x \cos s + \delta i_y \sin s)$$

$$d = \sqrt{\delta r^2 + (A\delta\theta)^2}$$

The rotation of one spacecraft around the other in the meridian plane is performed along the same type of curves as shown in Figure 2.3.E. At the two orbital plane crossings, where the direction from the Earth's centre is orthogonal to $\delta\bar{i}$, the two spacecraft fly alternatively above and below each other with the radial separation

$$\delta r = \pm A(\delta e_x \delta i_y - \delta e_y \delta i_x)/|\delta\bar{i}|$$

The closest approach, in the projection on the meridian plane, is smaller than the expression above. It can be calculated by the methods of the theory of quadratic forms to be:

$$d_{min} = \underset{s}{min}\, d = A\sqrt{B_1 - \sqrt{B_1^2 - B_2^2}} \quad \text{where}$$

$$B_1 = 0.5(|\delta\bar{e}|^2 + |\delta\bar{i}|^2) \quad ; \quad B_2 = \delta e_x \delta i_y - \delta e_y \delta i_x$$

The necessary and sufficient condition for obtaining $d_{min} \neq 0$ to ensure separation is:

$$\delta e_x \delta i_y - \delta e_y \delta i_x \neq 0$$

The left hand side is the scalar product of $\delta\bar{e}$ with a vector that is orthogonal to $\delta\bar{i}$. The criterion is that $\delta\bar{i}$ and $\delta\bar{e}$ are not zero and not parallel or antiparallel. The condition above is now the target separation of the orbital elements that shall be fulfilled in addition to the target conditions of mode 4.

In this co-location mode every spacecraft pair must have $\delta\bar{e} \neq 0$ like in Mode 4, so it does not provide any more co-location possibilities. Its only, but important, advantage is that it is independent of $\delta\lambda$ and thus removes the disadvantage of being sensitive to errors in the longitude and its drift rate. This makes it one of the most favoured co-location methods for missions with a low or medium accuracy tracking system that only provide, e.g., single station range and antenna angles, Chapter 8. The disadvantage is that the inclination manoeuvres must be coordinated between the members, which may be difficult if they have different manoeuvre constraints.

The best way to ensure that the separation criterion $\delta e_x \delta i_y - \delta e_y \delta i_x \neq 0$ in this mode is satisfied is to separate the eccentricity vectors as already described in mode 4. The same separation can then be copied for the plane of the inclination vectors but rotated by $+90°$ or $-90°$. It can be expressed as follows, where $C \neq 0$ is a positive or negative constant:

$$\delta i_x = -C\delta e_y \quad ; \quad \delta i_y = C\delta e_x$$

Figure G shows an arrangement of the inclination vectors for 4 co-located spacecraft that can be combined with either one of the two methods to separate the eccentricity vectors at the bottom of Figure F. The left of Figure G shows the 4 target inclination vectors. All the difference vectors $\delta\bar{i}$ stay approximately constant during the free drift until spacecraft numbers 3 and 4 reach the inclination boundary, to the right in Figure G. At this moment, the station keeping manoeuvres are performed for all members on the same day at the same time with the same $\Delta\bar{i}$ back to the targets to the left.

A small variation in the size and direction (i.e. execution time) of the different $\Delta\bar{i}$ will be necessary to correct for execution errors at the preceding station keeping manoeuvre. It is possible to let one, but only one, of the members follow the optimal long-term strategy of Section 6.4. In Figure G, spacecraft number 2 is most suitable. The other members should then keep its fixed off-set $\delta\bar{i}$ to the target of number 2.

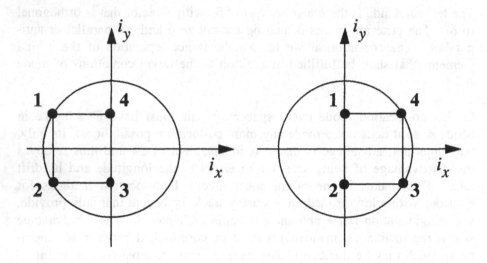

Figure 5.6.G. The plane of inclination vectors showing, left, start of the station keeping cycle with 4 separate targets. Right: End of the cycle. The manoeuvres are preformed at the same day and time by moving all the 4 inclination vectors back to the targets shown to the left. The circle in each plot indicates the maximum allowed inclination, with radius i_{max}.

When there are only two co-located spacecraft, they can both use the whole inclination deadband in spite of being separated. An inclination station keeping manoeuvre is performed by one spacecraft at the time, when the other one has drifted approximately to the middle of the deadband as shown in Figure H. One can have $\delta i_y = 0$ but must keep $\delta i_x \neq 0$ although its sign changes at each manoeuvre. It shall be combined with a separation in the eccentricity vector plane such that $\delta e_y \neq 0$ according to either one of the two methods shown at the top of Figure F. The jump of one inclination vector across the other at the manoeuvres does not cause any proximity risk since the thrust is applied near the orbital plane crossing, where there is a radial separation. Afterwards follows an out-of-plane separation with the inverse sign to the preceding part of the station keeping cycle.

Mode 6. General inclination and eccentricity separation.
The use of all three dimensions for separation does not increase the number of co-location possibilities, since each spacecraft pair must still have $\delta \bar{e} \neq 0$ as in mode 4. If we put $\delta \lambda_0 = \delta D = 0$ we obtain after some lengthy calculations the minimum separation in three dimensions:

$$d_{\min} = \min_{s} d = A \sqrt{B_1 - \sqrt{B_1^2 - B_2^2}} \quad \text{where}$$

$$B_1 = 2.5 \left| \delta \bar{e} \right|^2 + 0.5 \left| \delta \bar{i} \right|^2$$

$$B_2^2 = (\delta e_x \delta i_y - \delta e_y \delta i_x)^2 + 4 \left| \delta \bar{e} \right|^4 + 4 (\delta \bar{e} \cdot \delta \bar{i})^2$$

The above expression for B_1 contains two terms with contributions from, respectively, the in-plane and out-of-plane dimensions. In B_2, the three terms on the right hand side can be identified as the contributions from the separation in, respectively, the meridian plane of Mode 5, the equator plane of Mode 4 and a new inclination and eccentricity separation mode in the local horizontal plane. The latter mode provides an along-track separation at the orbital plane crossings, but it suffers from the same sensitivity to small errors in the longitude drift rate as Mode 4, and does not contribute to the number of co-location possibilities. As in mode 4, the necessary and sufficient criterion to get

$d_{\min} > 0$ is to have $\delta \bar{e} \neq 0$

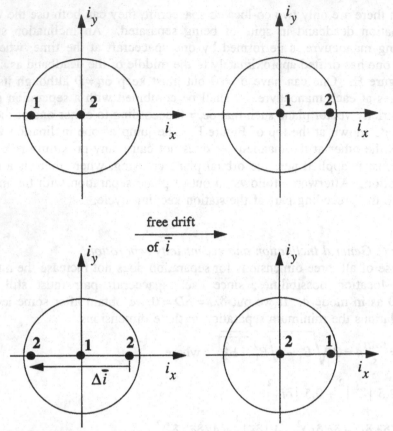

Figure 5.6.H. Two spacecraft can both use the whole inclination deadband (a circle with radius i_{max}) if the station keeping is out of phase by half a cycle, according to the following sequence.

- Top left: Begin of cycle for spacecraft 1; mid-point of cycle for spacecraft 2.

- Top right: Mid-point of cycle for spacecraft 1; end of cycle for spacecraft 2.

- Bottom left: Manoeuvre for spacecraft 2 by which $\bar{i}_2$ is moved by $\Delta \bar{i} \approx (-2i_{max}, 0)$; mid-point of cycle for spacecraft 1.

- Bottom right: Mid-point of cycle for spacecraft 2; end of cycle for spacecraft 1.

5.7 Proximity Manoeuvres

In this section we will deal with two different problems of great importance for planning manoeuvres during co-location. Since they both use the same mathematical method, we will treat them together here.

- The first question is how one can plan the normal station keeping or other orbital manoeuvres without risking any dangerous proximity in view of the unavoidable manoeuvre errors.

- The second question is how to plan a proximity avoidance manoeuvre if one predicts a dangerous close approach in the near future.

Starting with the first question, concerning thrust errors, one can consider two different ways to schedule manoeuvres. Avoidance manoeuvres should only use the first one, whereas regular manoeuvres can be performed either way.

- The usual way is to perform one thrust for one spacecraft at a time, then determine its orbit and check that the predicted orbit is safe from any of the co-located spacecraft before any new manoeuvre is made for any of them.

- Another way is to execute the station keeping manoeuvres at the same time for the co-located spacecraft.

The latter method is only advisable if one is confident that the combined thruster performance is good enough to ensure sufficient separation even in case of maximum dispersion, at least until the new orbits are determined and corrective action becomes possible. It is important in this mode that the manoeuvres really are performed close in time, since manoeuvring several spacecraft at different times but without the possibility to check the orbits in between provides the highest uncertainty and the greatest proximity risk.

In the mathematical formulas in this section we will denote the nominal predicted thrusts for the two spacecraft by $\Delta\overline{V}_1$ and $\Delta\overline{V}_2$, respectively, when they are manoeuvred at the same time $= t_b$. When only one of them is manoeuvred we can use the same formulas, after setting $\Delta\overline{V}_2 = 0$. We denote now by $\overline{r}_k(t)$, with $k = 1,2$, the two numerically integrated spacecraft orbits, including the effect of $\Delta\overline{V}_k$.

Fortunately one can neglect some types of errors in the prediction of thruster performance. The time t_b of execution can be considered as being absolutely predictable, for all practical purposes. The only other error to be considered is:

- For a longitude thrust; the size of the $\Delta \overline{V}$, whereas direction errors can be neglected.

- For inclination thrusts, the along-track component of the in-plane coupling according to Section 6.5.

The mathematical treatment of the errors is now simplified by the fact that, for both the above types of manoeuvres, the error in $\Delta \overline{V}$ to be considered is tangential. We then denote, for the two cases, the tangential component of the corresponding manoeuvre error by $\delta \Delta V_k$ for $k = 1,2$. When only one spacecraft is manoeuvred we put of course $\delta \Delta V_2 = 0$.

According to Section 3.3 we now get the linearised effect in the radial and tangential directions, respectively, of the tangential error on the orbits, relative to the error-free propagation $\bar{r}_k(t)$:

$$\delta r_k = 2(\delta \Delta V_k / \psi)[1 - \cos(s - s_b)]$$

$$A \delta \lambda_k = (\delta \Delta V_k / \psi)[4 \sin(s - s_b) - 3(s - s_b)]$$

By inserting the unit vectors in the radial and tangential directions we obtain the new expression for the orbital propagation in the inertial co-ordinate system, including the effect of the tangential errors:

$$\bar{r}_k(t) + \delta \bar{r}_k(t) = \bar{r}_k(t) + \delta r_k(\cos s, \sin s, 0) + A \delta \lambda_k(-\sin s, \cos s, 0)$$

Here we have inserted the sidereal angle s corresponding to the co-location mid-point longitude as the independent variable. In order to simplify the appearance of the subsequent equations we define now temporarily the time-dependent vector function:

$$\bar{q}(t) = (2/\psi)[1 - \cos(s - s_b)](\cos s, \sin s, 0) +$$

$$+ (1/\psi)[4 \sin(s - s_b) - 3(s - s_b)](-\sin s, \cos s, 0)$$

The previous expression for the orbit, including the effect of the tangential errors, becomes:

$$\bar{r}_k(t) + \delta \bar{r}_k(t) = \bar{r}_k(t) + \bar{q}(t) \delta \Delta V_k \quad \text{for } k = 1,2$$

In order to obtain the intersatellite distance, including the effect of the errors, we subtract the two equations for $k = 1$ and 2 from each others, after introducing:

$$\delta V = \delta \Delta V_1 - \delta \Delta V_2$$

$$\bar{r}_1(t) + \delta \bar{r}_1(t) - \bar{r}_2(t) - \delta \bar{r}_2(t) = \bar{r}_1(t) - \bar{r}_2(t) + \bar{q}(t)\, \delta V$$

We get the total distance d from the square of the absolute value of the vector expression above:

$$d^2 = \left| \bar{r}_1(t) - \bar{r}_2(t) + \bar{q}(t)\, \delta V \right|^2 = C_0(t) + C_1(t)\delta V + C_2(t)\delta V^2$$

The right hand side shows the spacecraft distance expressed as a quadratic form in δV with time-dependent coefficients:

$$C_0(t) = \left| \bar{r}_1(t) - \bar{r}_2(t) \right|^2$$

$$C_1(t) = 2[\bar{r}_1(t) - \bar{r}_2(t)] \cdot \bar{q}(t)$$

$$C_2(t) = \left| \bar{q}(t) \right|^2$$

These coefficients are known from the accurate numerical integration of the two $\bar{r}_k(t)$ and the simple analytical function $\bar{q}(t)$. It is worth mentioning here that the only approximations made so far concerns the influence of the manoeuvre errors on the orbit. This is not any major disadvantage since the errors are, in any case, not very well known. The condition that the distance shall not be less than a minimum safety distance $d_{\min}$ becomes now:

$$C_0(t) + C_1(t)\delta V + C_2(t)\delta V^2 \ge d_{\min}^2 \quad \text{when } t > t_b$$

The upper limit of the time interval can, for practical purposes, be set to the time it would take to determine the new orbit or orbits and take some corrective action. What is done in practice for the manoeuvre planning is that one calculates for each time (step-wise with interpolation between the steps) the coefficients of the quadratic equation below and solves the two roots in δV:

$$C_2(t)\delta V^2 + C_1(t)\delta V + C_0(t) - d_{\min}^2 = 0$$

If there is no real root δV there is no risk for proximity violation for the corresponding time, regardless of the size of the tangential manoeuvre error. When there are two real roots, they define the end-points of the forbidden interval in the error δV. If an error within this interval is physically possible from what is known of the thruster dispersion, one should change the manoeuvre planning, or risk a spacecraft approach below d_{min}.

When considering the size of the possible δV dispersion it is advisable to use an unsymmetrical interval. An over-performance is unlikely to be more than up to 20% for any thrusting system. For the under-performance one must normally consider all values down to any $\Delta V_k = 0$ to account for the possibility that the manoeuvre may be postponed or interrupted.

A proximity avoidance manoeuvre can also be planned by means of the same mathematics as already described. We denote here by $\bar{r}_k(t)$ the two nominal predicted spacecraft trajectories without manoeuvres or with nominally planned manoeuvres. If we then see that the distance for some time in the future will be

$$|\bar{r}_1(t) - \bar{r}_2(t)| < d_{min}$$

we can avoid the proximity by performing a tangential thrust ΔV_1 at the time t_b with spacecraft number 1. This results in the new predicted distance, squared:

$$d^2 = |\bar{r}_1(t) + \bar{q}(t)\,\Delta V_1 - \bar{r}_2(t)|^2 = C_0(t) + C_1(t)\Delta V_1 + C_2(t)\Delta V_1^2$$

with the same definitions as before. For the times around the close approach we will then obtain one positive and one negative real root when solving ΔV_1 from the equation

$$C_0(t) + C_1(t)\Delta V_1 + C_2(t)\Delta V_1^2 = d_{min}^2$$

The proximity avoidance manoeuvre can be an east or a west thrust with a ΔV_1 outside the interval between the two roots. In order not to introduce any new proximity it should also lie outside the union of the forbidden intervals of the solutions of the quadratic equations for all other times t in the foreseeable future. This may seem to be a difficult constraint to satisfy, but one still has some freedom to choose the thrust time t_b by trying out different times between the present and the proximity.

It is recommended to make the avoidance manoeuvre at least half a day before the feared proximity violation. A thrust of 1 m/s can cause, after half a day, an increase in separation distance of 54.8 km in radial and 130 km in tangential direction according to Section 3.3. Figure A shows the total distance = $|\bar{q}(t)|$ multiplied by a $\Delta V_1 = 1$ m/s.

It is also possible, although not very fuel efficient, to make an out-of-plane thrust for proximity avoidance. The optimal time is then six hours before the predicted proximity violation. According to Section 3.2, a thrust of 1 m/s can increase the latitude separation by 13.7 km.

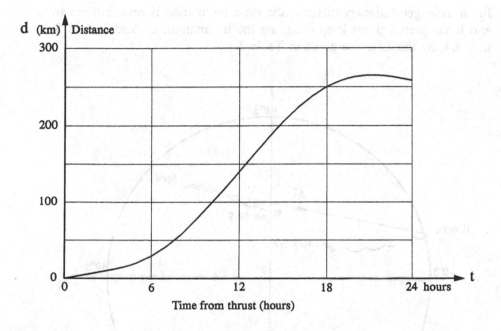

Figure 5.7.A. In-plane distance during the first 24 hours after a 1 m/s tangential thrust.

6. INCLINATION STATION KEEPING

6.1 Principles of Station Keeping

In an ideal geostationary mission, the orbit inclination is zero, but the solar and lunar perturbations keep changing the inclination, as described in Section 4.4, by the drift rate given in Table 3.

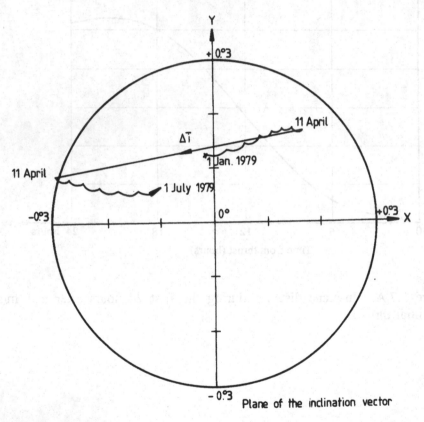

Figure 6.1.A. Inclination manoeuvre for ESA's Meteosat-1 on 1979 April 11 at 17:17 UTC. The sidereal angle was 100.2° and ΔV = -25.19 m/s (south thrust) including a 7% overshoot. The circle shows the constraint $i_{max} = 0.3°$ that was applied for this part of the mission.

The result of a non-zero inclination is a libration in spacecraft latitude between $+i$ and $-i$ with the period of one sidereal day, Section 2.3. The mission requirements usually prescribe a maximum allowable inclination, i_{max}, depending on the size of the latitude librations that can be tolerated. Typical values of i_{max} range from about 0.05° up to several degrees.

We will use here the inclination vector $\bar{i}$ interleaved with (i, Ω) for the two out-of-plane orbit parameters. The inclination constraint

$$i = |\bar{i}| = \sqrt{i_x^2 + i_y^2} \leq i_{max}$$

is then represented by a circle in the $\bar{i}$-plane, Figure A. For convenience, the $\bar{i}$-vector in this figure is expressed in degrees rather than in radians.

In Section 4.4 we showed that the perturbing forces act on the orbit inclination with a north acceleration when the spacecraft is near the +y-axis and a south acceleration near the -y-axis. The mean direction of the drift vector $d\bar{i}/dt$ is then approximately in the +x-direction, Figure B. To compensate for the perturbations, the inclination station keeping manoeuvres then have to be given as south thrusts near the +y-axis, i.e. at a sidereal angle near 90° or as north thrusts when the spacecraft is near a sidereal angle of 270°. Since the sidereal angle of the thrust is approximately given, the time of day of the thrust must be selected according to the time of the year. The local time, at the spacecraft longitude, is given by the following table.

	North Thrust	*South Thrust*
Spring	Morning	Evening
Summer	Midnight	Noon
Autumn	Evening	Morning
Winter	Noon	Midnight

At these sidereal angles there is no risk of an eclipse by the Earth during or near the manoeuvre. There may, however, be constraints from the spacecraft design that forbids manoeuvres when the angle between the Earth and Sun, projected on the orbital plane, lies below a given value. The reason could be that the attitude control system on board relies upon independent reference directions to the Earth and Sun during the thrusting. Another reason can be risk of damage to the solar arrays by the thrust exhaust at certain array angles relative to the spacecraft body. In either case the consequence may be that one has to refrain from inclination manoeuvres during a certain time around midsummer and midwinter, or one must manoeuvre the inclination vector in a non-optimal direction.

The usual way of performing inclination station keeping is to let $\bar{i}$ drift, as in Figure A, until it comes close to the boundary circle with radius i_{max}. One then performs a manoeuvre to move $\bar{i}$ by a $\Delta\bar{i}$, the size of which is nearly $2i_{max}$ to the opposite side of the circle. The result of an impulsive thrust ΔV appears like an instantaneous jump $\Delta\bar{i}$ of the inclination vector in Figure A. The size of the jump, $|\Delta\bar{i}|$, is proportional to ΔV by 1° for 53.7 m/s.

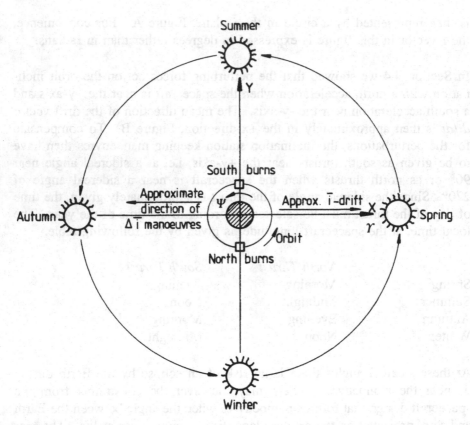

Figure 6.1.B. Station keeping south thrusts are performed near the +y-axis and north thrusts near the -y-axis. The time of day of the thrust depends on the time of the year.

The approximate time between manoeuvres can be obtained by comparing the size of the deadband $2i_{max}$ with the natural inclination drift rate from Table 3. With $i_{max} = 0.5°$, one manoeuvre per year is needed, whereas larger i_{max} leave correspondingly longer times between manoeuvres. With smaller values of i_{max}, however, one must manoeuvre more frequently than according to what is obtained by the simple calculation above, since the nonlinear drift of $\bar{i}$ then becomes relatively more important.

The time of day of the thrust and on which day the thrust is executed are two independent parameters in the planning of an inclination manoeuvre. The expression "time of the manoeuvre" will in the following be used only for the latter, whereas the former is indirectly referred to through the direction of $\Delta \bar{i}$. The direction and size of the inclination station keeping manoeuvres are normally calculated by means of an optimisation program according to the principles described in the subsequent sections.

Because of the high fuel consumption needed for inclination station keeping, compared with longitude station keeping, it is particularly important to optimise the inclination manoeuvres. Three typical optimisation tasks can be distinguished:

- Minimise the fuel needed to keep the inclination below i_{max} for a given mission lifetime;

- Maximise the lifetime during which the inclination can be kept below i_{max} with a given amount of fuel on board the spacecraft;

- Minimise the maximum inclination during a given lifetime with a given amount of fuel on board.

Usually, however, the conditions of optimisation are not uniquely defined. Instead, one often has a given amount of fuel on board that has been calculated before the launch to be able to keep the inclination below i_{max} for at least a given mission lifetime. The fuel left after the end of the planned mission time is then often used to prolong the mission, so in practice the second case above is applicable. Other types of inclination constraints also exist e.g. that different values of i_{max} apply to different parts of the mission. A special case of the latter is that often, towards the end of a mission when fuel is getting short, the mission is prolonged by abandoning the inclination station keeping completely.

These three optimisation criteria for the fuel only define the direction of the $\Delta\bar{i}$ manoeuvres, but not the timing and size of each manoeuvre. With the given direction of $\Delta\bar{i}$ one can select the remaining parameters such that one of the three criteria applies:

- Maximise the time until the next manoeuvre is needed. This means moving $\bar{i}$ to the left, as in Figure A, opposite to the mean drift direction, until it reaches the periphery of the circle $|\bar{i}| = i_{max}$. This leaves a long free drift time until $\bar{i}$ drifts to the right-hand boundary of the circle again. From the size of $|\Delta\bar{i}|$ one obtains the thrust size ΔV.

- Perform the manoeuvres according to a fixed schedule, e.g. once every 14 days, but each time move $\bar{i}$ as above. This usually means that each manoeuvre is executed before $\bar{i}$ reaches the right-hand part of the circle $|\bar{i}| = i_{max}$ but it gives a certain margin to avoid violating $|\bar{i}| = i_{max}$ in case of delay or failure of the next manoeuvre.

- Perform the manoeuvres on a regular schedule as above, but choose the thrust sizes such that the maximum value of i is as small as possible in the interval between each manoeuvre.

The selection of which station keeping criterion to apply depends on the requirements of the mission. The decision should be taken during the mission planning phase before the launch. In many cases, however, it is not spelled out explicitly in the mission requirement document, so the spacecraft Operator has to decide which station keeping mode is best suited.

6.2 Passive Inclination Control

Passive inclination control means that no inclination manoeuvre is needed during the routine operation of the mission. This is possible if the allowed inclination i_{max} is so large and the mission lifetime is so short that the total inclination drift during the life is smaller than $2i_{max}$. By adding up the annual drifts from Table 3 for consecutive years one can roughly calculate for how long one can passively control the inclination below a given i_{max}.

The optimum starting condition of the inclination vector can be calculated by a series of numerical integrations of the inclination vector drift. Figure

A shows as an example calculations for a 5-year mission, from January 1981 to December 1985, with $i_{max} = 2.5°$. We integrate $\bar{i}$ backwards in time, starting at the mission end with the initial (= end) condition $i = i_{max}$ and a few different values of Ω. Depending upon Ω, some of the resulting curves of $\bar{i}$ lie inside the circle $i < i_{max}$, whereas in other cases the left-hand part of the curve lies outside.

With a very good approximation, one can fit a circle of 2.5° radius through the left-hand end of the curves, corresponding to the beginning of the mission. The region between the two circles, shaded in Figure A, is then the margin inside which the inclination vector must lie at the beginning of the mission in order to satisfy the inclination criterion for the whole mission duration. The size of the starting region depends on i_{max} and the lifetime. For some missions it may be wide because of relaxed criteria, whereas for others it may shrink to only a point in the $\bar{i}$ plane. In all cases, the starting point will be near or somewhat below the -x-axis, corresponding to an initial value of Ω near or somewhat higher than 270°.

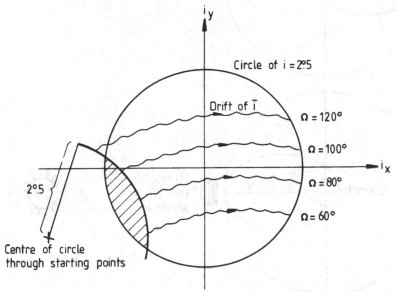

Figure 6.2.A. Five years passive inclination control, from January 1981 to December 1985, with $i_{max} = 2.5°$. The two-week waves in the inclination vector drift are smoothed out, whereas the half-year waves are clearly seen. The inclination vector must lie inside the shaded area at the beginning of the mission in order to satisfy $|\bar{i}| \leq i_{max}$ for five years.

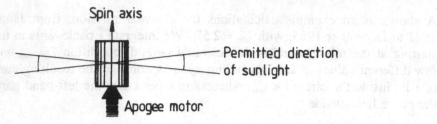

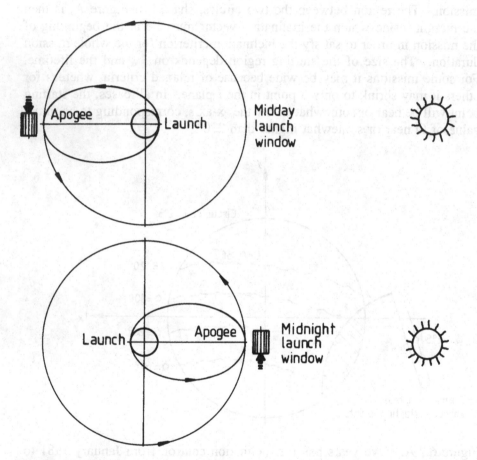

Figure 6.2.B. The transfer orbit apogee must point away from or towards the Sun in order to ensure that the spin axis at apogee motor firing is approximately orthogonal to the Sun direction.

In order to give a complete picture of passive inclination control we will mention briefly how to obtain the required starting conditions for $\bar{i}$ at the injection from the transfer orbit (Section 1.1) to the geostationary orbit. It is more difficult to launch a satellite to obtain a starting inclination of several degrees and a specified ascending node than to start with the inclination near zero. In the former case one must perform a *node rotation*, which is the change in node between the transfer orbit and the geostationary orbit. The difficulty is due to the dependence of the node rotation on the launch date, resulting from a long chain of interconnected requirements, as follows.

In order to minimise the amount of fuel needed for station acquisition one tries to define a transfer orbit (Section 1.1), the apogee of which touches the target geostationary orbit in a point where the apogee motor can be fired, Figure B. The latitude along this desired geostationary orbit, with a non-zero inclination, depends on the sidereal angle, Figure C. Unfortunately, however, one has not the freedom to select neither the sidereal angle, nor the latitude of the transfer orbit apogee, as explained below. The discrepancy has to be compensated for by the station acquisition manoeuvres.

The sidereal angle of the transfer orbit apogee is constrained by the *launch window*. It defines the daily time interval allowed for launch and is usually a midday and/or midnight window, meaning that the direction from Earth to the transfer orbit apogee is approximately aligned with the Earth/Sun direction, Figure B. The reason is that on a spin stabilised spacecraft the spin axis, which is aligned with the motor firing direction, must stay orthogonal to the Sun direction with a certain margin, for thermal reasons. A similar constraint may apply to a three-axis stabilised spacecraft, where a certain Sun-Earth geometry in a given part of the transfer orbit is imposed by the attitude control system.

It is highly desirable to be able to launch at any day of the year in order to accommodate possible delays in spacecraft or launcher availability. The consequence is that the desired latitude at the apogee motor firing can have any value inside $\pm i_{max}$, as seen in Figure C. The transfer orbit elements are obtained from the launch trajectory that is programmed in the launcher on a *launch tape*. Each launch tape is expensive and must be programmed long before the launch, so it is not possible to adapt the transfer orbit latitude to the launch day when the precise day of launch cannot be predicted far in advance.

A further difficulty applies to apogee motors with solid fuel that can only
produce a fixed thrust by the fixed amount of fuel that is loaded before the
motor is integrated into the spacecraft. When the motor burns liquid fuel
one has got more flexibility to adapt the injection thrust by ground command to the size that is optimal for the current node rotation. However, it
is not always possible to make use of the multiple thrust injection to fire
some of the thrust away from the apogee where the latitude may be more
favourable for node rotation because there may be constraints from other
sources.

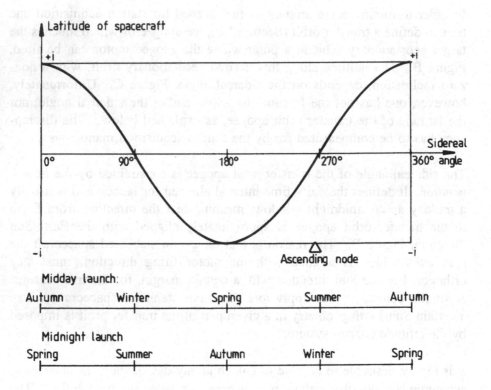

Figure 6.2.C. The latitude of a geostationary spacecraft with non-zero
i and $\Omega = 270°$ varies as the cosine of the sidereal angle between
$+i$ and $-i$. The apogee motor should preferably be fired at the latitude
of the geostationary orbit, which depends on the sidereal angle of the
transfer orbit apogee. This sidereal angle, in its turn, depends on the time
of year of the launch.

The solution to the conflicting constraints is a compromise with one or a few different launch tapes, possibly combined with restricting the launch to only part of a year. The tape or tapes are made to match, as good as possible, the mean requirements for one launch slot of the year. As a result of the compromise, combined with the launcher and firing errors, the initial value of $\bar{i}$ after the apogee motor firing may or may not lie inside the desired starting region. If it lies outside, one can move it inside by a station acquisition inclination manoeuvre.

If, however, $\bar{i}$ initially lies inside the circle $i \leq i_{\max}$, but outside the optimal starting region, one does not need to perform the inclination manoeuvre immediately, but can wait until $\bar{i}$ approaches the boundary $i_{\max}$. It is only a matter of definition whether this is called a station acquisition or a station keeping manoeuvre. The fuel consumption is the same and the effect on $\bar{i}$ is the same in both cases. The next sections describe how the size and direction are calculated for such station keeping or station acquisition manoeuvres.

6.3 Active Inclination Control

Let us assume that we have a geostationary orbit with the inclination vector $\bar{i}_0$ at the time now $= t_0$ in Figure A. By integrating the drift of $\bar{i}$ forwards in time we see that it will not satisfy the constraint $i \leq i_{\max}$ at the end of the mission t_1.

The question of how to perform a manoeuvre now to ensure that $i \leq i_{\max}$ during the remaining part of the mission $t_0 \leq t \leq t_1$ can be solved in the same way as shown in Figure 6.2.A in the previous section. We integrate $\bar{i}$ backwards from t_1 to t_0 to obtain the permissible target region, the shaded area in Figure A. The fuel minimum manoeuvre $\Delta \bar{i}$ from $\bar{i}_0$ to the target region is found easily by geometric means. The direction of $\Delta \bar{i}$ coincides with the direction from $\bar{i}_0$ to the centre "X" of the left-hand circle in Figure A.

A slightly different method is used to calculate the manoeuvre direction when one has a given amount of fuel allocated for an inclination thrust Δi. Figure B shows the circle of attainable values of $\bar{i}$ centred around the present value $\bar{i}_0$. To optimise the thrust $\Delta \bar{i}$ one then integrates the drift of $\bar{i}$ forward in time, starting at t_0 from different initial values on this circle. If the optimisation criterion is to stay inside $i \le i_{max}$ for as long as possible, one interrupts each integration when reaching the right-hand circle $i = i_{max}$ in Figure B. The starting value giving the longest mission time is then the manoeuvre target.

A method analogous to that shown in Figure B is used when the criterion is to minimise the maximum i for a given mission time $t_0 \le t \le t_1$. Instead of interrupting the integration at a given i one continues each drift integration until the mission end t_1. The curve giving the lowest value of i is then selected for the manoeuvre target.

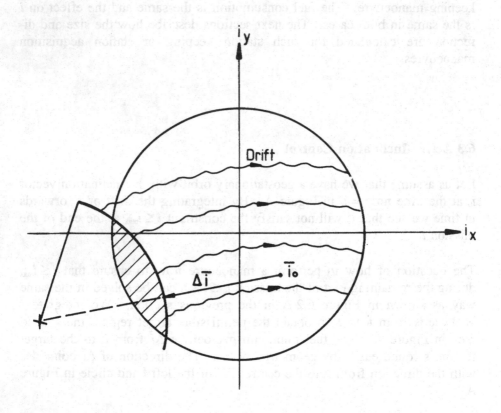

Figure 6.3.A. $\Delta \bar{i}$ is the fuel-minimum station keeping manoeuvre that makes $\bar{i}$ stay inside the given circle for a fixed mission duration.

The optimisations described here assume that the day of the manoeuvre t_0 is fixed. The question then arises what day to select if one knows that the inclination will drift outside i_{max} at some future date, before the end of the mission. Should one perform the manoeuvre immediately or wait until just before the inclination drifts outside?

The answer is that it does not really matter, as far as the fuel consumption is concerned, with a very good approximation. This approximation means essentially that the natural drift of $\bar{i}$ is assumed to be the same in different parts of the $\bar{i}$-plane, apart from a rotation around the zero point. The error in this approximation is proportional to the inclination and amounts to less than 0.01 deg/year when $i = 1°$. In most practical cases this approximation is sufficient for manoeuvre planning, and the times of the inclination manoeuvres are planned according to operational convenience instead of fuel-consumption optimisation.

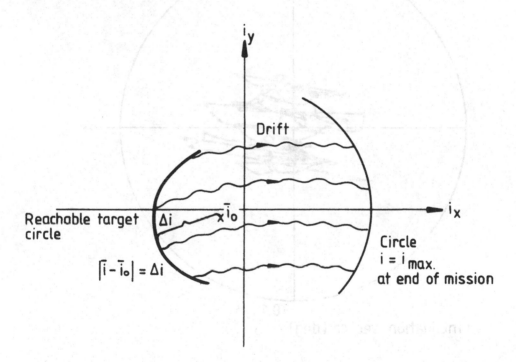

Figure 6.3.B. Optimisation of the direction of an inclination thrust with given size Δi. The resulting inclination must stay below i_{max} for as long as possible.

When the allowed inclination deadband i_{max} is small, one has to perform frequent inclination station keeping manoeuvres to compensate for the drift of $\bar{i}$. Figure C shows the trajectory of $\bar{i}$ during one year with 14 manoeuvres for ESA's OTS spacecraft. The inclination deadband was nominally defined to be 0.1° for OTS, but in practice the inclination was kept below 0.07° in this time period.

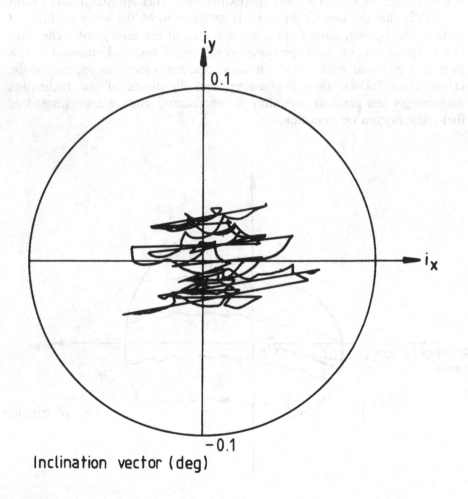

Figure 6.3.C. Inclination vector of ESA's Orbital Test Satellite (OTS) from mid-1980 to mid-1981 with 14 inclination manoeuvres. The inclination was kept well below the originally defined $i_{max} = 0.1°$.

A better visibility of the manoeuvres is obtained by plotting the $\bar{i}$-plane as in the theoretical example of Figure D. Here $i_{max} = 0.1°$ and the manoeuvres are performed in a constant direction by a $\Delta\bar{i}$ of magnitude 0.1° on a regular time schedule. In Figure D the origin of the co-ordinate system, including the circle $|\bar{i}| = i_{max}$, is shifted by $-\Delta\bar{i}$ to the right at each manoeuvre $\Delta\bar{i}$ instead of the usual representation where $\bar{i}$ is moved by $+\Delta\bar{i}$ to the left. One can see that it is possible to manoeuvre $\bar{i}$ along a straight line only if the inclination circle is wide enough to accommodate the wavy motion of the y-component i_y.

Inclination manoeuvres can be planned by manoeuvre optimisation programs of various degrees of sophistication. Simple methods like those shown in Figures C and D can be prepared by some hand calculations with the aid of the general orbit prediction program. Such station keeping, however, does not always give completely optimal fuel consumption. A high degree of optimisation is provided by the method described in the next section. This type of fuel optimisation only calculates the direction of $\Delta\bar{i}$, whereas the day of the manoeuvre and the size of each $\Delta\bar{i}$ do not influence the total amount of station keeping fuel during the mission lifetime.

The preliminary estimate of the annual ΔV given in Table 3 assumes that the station keeping manoeuvres move only the x-component i_x of $\bar{i}$ back to zero at the end of each year, but lets the y-component i_y drift free. However, when the inclination deadband i_{max} is small, the station keeping ΔV is larger, for two reasons.

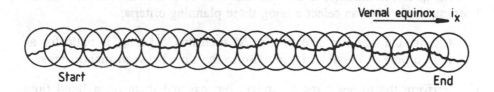

Figure 6.3.D. Theoretical calculation of station keeping manoeuvres during 3 years. In this figure the manoeuvres are shown by shifting the circle $|\bar{i}| = i_{max} = 0.1°$ to the right at each manoeuvre instead of moving $\bar{i}$ to the left, as in Figure C.

One reason is that the usual saving, of a thrust equivalent to $2i_{max}$, becomes smaller. This saving, at the beginning and end of the mission, is obtained by starting with $\bar{i}$ to the left of the circle $|\bar{i}| = i_{max}$ and ending at the right, as in the case of passive inclination control.

The second reason is due to the direction of the $\Delta\bar{i}$ as follows. When i_{max} is very small, say below 0.03° one must move $-\Delta\bar{i}$ along a zig-zag line as in Figure E in order to keep the waves of the drift motion of $\bar{i}$ inside the small circle $|\bar{i}| = i_{max}$. This type of station keeping costs more fuel than moving in a constant direction, according to the theorem that a zig-zag line is not the shortest path between two points. Even smaller values of i_{max} would make it necessary to follow also the small two-week-period waves caused by the Moon.

However, when there is some margin inside the deadband i_{max} one can make use of the fact that the drift of i_x is lower when i_y is more negative. This appears from the approximate description of the drift given in Section 4.4 as a negative rotation in the $\bar{i}$-plane around the point $(0, -7.4°)$. A manoeuvre optimisation algorithm based upon this principle is given in the next section. However, when manoeuvring $\bar{i}$ at the lower part of the -y-axis one must let the drift curve of $\bar{i}$ stay sufficiently deep inside the boundary circle $|\bar{i}| = i_{max}$ (Figure F) in order to prevent the free drift length from becoming too short. The size of this depth depends on other mission constraints, but normally one needs a free drift time of at least one week between inclination manoeuvres.

When i_{max} is very small, as in Figure E, the direction of $\Delta\bar{i}$ has to be chosen at each thrust, in addition to its size and time of execution. In this station keeping mode one can select among three planning criteria:

• Choose the size and direction of $\Delta\bar{i}$ such that $|\bar{i}| \leq i_{max}$ is valid for as long a time as possible, i.e. maximise the time between manoeuvres.

• Perform the manoeuvres as above, but execute them on a fixed time schedule, e.g. every 7 days, to leave an inclination margin at the end of each cycle.

• Perform the manoeuvres on a regular schedule, as in the previous mode, but choose the direction and size of each $\Delta\bar{i}$ such that the maximum value of i is as small as possible in the interval between each manoeuvre.

The exact selection of the applicable station keeping criterion is normally not specified in the mission requirements document, so the spacecraft Operator has to decide which mode to use in each concrete situation.

Start   End

Figure 6.3.E. Enlarged part of the inclination vector drift from Figure D. The circles represent a constraint of $i_{max} = 0.03°$. Since the circles are so small they have to follow the wavy motion of i with half a year's period. The resulting zig-zag manoeuvre direction shown by the sequence of circles costs more fuel than the straight manoeuvre motion of Figure D.

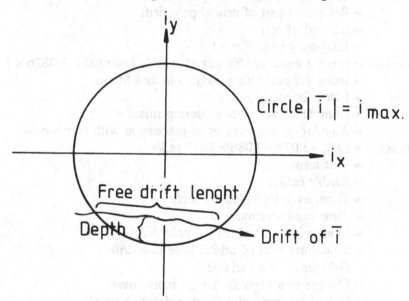

Figure 6.3.F. One must plan the long-term manoeuvre direction strategy such that the drift curve of $\bar{i}$ moves at a certain depth inside the circle $|\bar{i}| = i_{max}$. In this way one can avoid the free drift length of $\bar{i}$ becoming too short.

6.4 Long-term Strategy

This section contains more advanced mathematics than most other sections, but it can be omitted by readers who are not preparing inclination station keeping optimisation programs. It is copied from the paper "Geostationary Orbit Inclination Strategy" in the ESA Journal volume 9 of 1985, pages 65 to 74, after improving the approximation in Equation 15, which leads to a smaller remainder in Equation 17.

The following symbols are used in this section

$B_1(t), B_2(t)$ = Constraint functions for manoeuvre direction
C_1, C_2 = Coefficients in rest term
d = Margin of inclination deadband
$\bar{e}_z$ = Unit vector along the z-axis = (0,0,1)
F = Cost function for fuel optimisation
$h(t)$ = Rotational part of orbital pole drift
$H(t)$ = Integral of $h(t)$
j = Imaginary unit; $j^2 = -1$
$J_2 = -C_{20}$ = Lowest zonal coefficient of Earth's potential = 1.0826×10^{-3}
k = Index for perturbing body; Sun and Moon
m = Index for manoeuvres
M = Number of manoeuvres during mission
$\bar{p}$ = Angular velocity vector of precession with components:
(p_x, p_y, p_z) = (0.0, +3.079, -7.086)$\times 10^{-12}$ rad/s
Q = Rest term
R = Earth's radius
t = Time, as independent variable
T = Time until mission end
γ = Translation part of orbital pole drift
Γ = Transformation of orbital pole free drift
δ = Difference of a variable
Δ = Change in a variable from a manoeuvre
ζ = Complex representation of inclination vector
η = Transformation of ζ
ρ = Centre of deadband circle in the complex plane
τ = Integration variable for t
Φ = Free drift velocity in complex notation

The method used in this long-term manoeuvre strategy can be summarised in the following way. By means of a suitable transformation and approximation of the equation for the orbital pole drift one can reduce the optimisation of the directions of all future inclination manoeuvres to the task of finding the shortest path between two points = a straight line.

Inclination manoeuvres for geostationary orbits are expensive to perform, compared with longitudinal station keeping, and it is therefore worthwhile trying to optimise the strategy employed whenever possible. When the restriction on the maximum allowable orbital inclination is not too severe, at least 0.1°, one has a certain freedom to move the orbital pole inside the inclination deadband. This freedom can be used for finding a long-term strategy that minimises the fuel consumed for inclination station keeping throughout the mission.

The problem in devising a long-term strategy is that the planning for each manoeuvre must consider its consequences for all future manoeuvres, in addition to the natural pole drift. Consequently, some simplifications are needed in modelling both drift and future manoeuvres to establish the target parameters for an inclination manoeuvre. Once this target has been found, the orbit propagation itself is calculated with high accuracy by numerical integration of spacecraft position and velocity, taking all relevant accelerations into account.

For modelling the perturbing accelerations that cause the natural drift of the inclination vector we express, as before, the spacecraft position in terms of the vector $\bar{r} = (x,y,z)$ and its velocity by $\bar{V} = d\bar{r}/dt$ in the quasi-inertial Mean Equatorial Geocentric System of Date (MEGSD), which moves slowly as a function of time t from the precession of the mean equator and equinox. The linearised transformation from a vector $\bar{r}$ in MEGSD to a vector $\bar{r}_0$ in the inertial mean coordinate system at the epoch t_0 can be written as follows, under the assumption that t and t_0 are sufficiently close so that the angle of rotation is very small.

$$\bar{r}_0 = \bar{r} + (t - t_0)\,\bar{p} \times \bar{r} \tag{1}$$

For an orbit expressed in MEGSD the precession acts like a perturbation, of which we only need to include the Coriolis acceleration of about of 4×10^{-8} m/s^2. The perturbations from the nonspherical Earth potential is represented only by the lowest zonal term, from Section 4.2. Here we will represent the coefficient by the frequently used alternative notation $J_2 = -C_{20}$. In order to express the gradient of the corresponding spherical harmonic we need to define a unit vector $\bar{e}_z$ parallel to the positive z-axis.

The equations for the perturbations from the Sun and Moon are given in
Section 4.3. The total acceleration to be considered on the spacecraft now
becomes:

$$\frac{d^2\bar{r}}{dt^2} = -\frac{\mu}{r^3}\bar{r} + \sum_{k=1}^{2}\mu_k\left(\frac{\bar{r}_k - \bar{r}}{|\bar{r}_k - \bar{r}|^3} - \frac{\bar{r}_k}{r_k^3}\right) -$$

$$-\frac{3\mu J_2 R^2}{r^5}\left(z\bar{e}_z - \frac{5}{2}\frac{z^2}{r^2}\bar{r} + \frac{1}{2}\bar{r}\right) - 2\bar{p}\times\bar{V} \qquad (2)$$

In order to integrate the free drift of the orbital pole over the complete
mission duration of several years, we introduce a simplified model of the
longitudinal motion and the longitude station keeping. For this purpose,
we assume that the spacecraft moves along a perfect circle with radius equal
to the ideal geostationary semimajor axis A around the inclination vector $\bar{I}$
with the same angular velocity ψ as the Earth. If we count the time t from
an ascending node we can express the position and velocity of the spacecraft
as:

$$\bar{r} = A\cos\psi t\begin{pmatrix}\cos\Omega \\ \sin\Omega \\ 0\end{pmatrix} + A\sin\psi t\begin{pmatrix}-\sin\Omega\cos i \\ \cos\Omega\cos i \\ \sin i\end{pmatrix} \qquad (3)$$

$$\bar{V} = -A\psi\sin\psi t\begin{pmatrix}\cos\Omega \\ \sin\Omega \\ 0\end{pmatrix} + A\psi\cos\psi t\begin{pmatrix}-\sin\Omega\cos i \\ \cos\Omega\cos i \\ \sin i\end{pmatrix} \qquad (4)$$

When calculating the spacecraft velocity one considers Ω and i to be constant, since they change slowly compared with ψt. The angular momentum
vector of the orbital motion

$$\bar{r}\times\bar{V} = \psi A^2\bar{I} \qquad (5)$$

drifts under the influence of the perturbation part of the total acceleration,
which forms a couple with $\bar{r}$:

$$\psi A^2\frac{d\bar{I}}{dt} = \frac{d}{dt}(\bar{r}\times\bar{V}) = \bar{r}\times\frac{d^2\bar{r}}{dt^2} \qquad (6)$$

The right-hand side of Equation 6 is obtained by inserting Equation 2. After
the insertion, we approximate the expression for the Sun and Moon pertur-

bation with the same linearisation as used in Section 4.3. The differential equation for $\bar{I}$ becomes:

$$\frac{d\bar{I}}{dt} = \frac{1}{A^2\psi}\left[\sum_{k=1}^{2}\frac{3\mu_k}{r_k^5}(\bar{r}\cdot\bar{r}_k)(\bar{r}\times\bar{r}_k) - \frac{3\mu J_2 R^2 z}{r^5}\bar{r}\times\bar{e}_z - 2\bar{r}\times(\bar{p}\times\bar{V})\right] \quad (7)$$

Next, Equations 3 and 4 for $\bar{r}$ and $\bar{V}$ will be inserted on the right-hand side of Equation 7. Thereafter one can perform an analytical integration over one orbital period (= 1 sidereal day) = $\delta t = 2\pi/\psi$ while keeping the slowly varying vectors $\bar{I}$ and $\bar{r}_k$ constant during the integration. The resulting difference equation for the free drift of $\bar{I}$ now becomes

$$\frac{\delta\bar{I}}{\delta t} = \frac{1.5}{\psi}\left[\sum_{k=1}^{2}\frac{\mu_k}{r_k^5}(\bar{r}_k\cdot\bar{I})(\bar{r}_k\times\bar{I}) - \frac{\mu J_2 R^2}{A^5}(\bar{e}_z\cdot\bar{I})(\bar{e}_z\times\bar{I})\right] - \bar{p}\times\bar{I} \quad (8)$$

It is sufficient to project the equation on the x-y-plane and only calculate the drift of the two-dimensional inclination vector $\bar{i}$. As will become apparent later, it is an advantage to express it as a complex number ζ, where we denote the imaginary unit by j ($j^2 = -1$) since i is used for the inclination.

$$\zeta = i_x + j\,i_y = \sin i \exp[j(\Omega - \pi/2)] \quad (9)$$

The time step in the difference Equation 8 is one day, but this is very small in comparison with the long-term strategy, which requires the evolution of the inclination over several years. For this reason we can replace the left-hand differential by the derivative. By replacing the dependent variable from Equation 9, we obtain the mean drift in ζ in the form

$$\frac{d\zeta}{dt} = \Phi(\zeta, t) \quad (10)$$

From the right-hand side of Equation 8 we can express the complex-valued function Φ of the complex variable ζ and the real variable t. The time t enters into Φ only via the time-dependence of the Sun and Moon positions $\bar{r}_k$. In order to find sufficiently simple analytical expressions for the inclination drift and station keeping, we will now linearise the equations with respect to small values of the inclination, i.e. we put

$$|\zeta| = \sin i \approx i \quad (11)$$

The station keeping requirement to keep the inclination below a given maximum value i_{max} can be expressed as

$$|\zeta| \le i_{max} \tag{12}$$

The thrust for a north ($\Delta V > 0$) or south ($\Delta V < 0$) manoeuvre, performed when the spacecraft is at the sidereal angle s, changes the inclination vector, according to Section 3.2, by:

$$\zeta_2 - \zeta_1 = \Delta\zeta = (\Delta V/V)\exp[j(s - \pi/2)] \tag{13}$$

The fuel consumption of a manoeuvre is proportional to $|\Delta\zeta|$. The direction of the pole motion is equal to arg $\Delta\zeta$ and will be referred to as the manoeuvre direction. It depends on the time of day of the thrust and the spacecraft longitude. The expression "time of manoeuvre" will be used in the following to denote essentially on which day the manoeuvre is executed. The time within the day enters only into the calculation of the spacecraft sidereal angle, which influences the manoeuvre direction.

In the free-drift Equation 10 we can expand the function Φ in powers of the small variable ζ. For the zero order term we introduce the complex-valued function

$$\gamma(t) = \Phi(0,t) = \frac{1.5}{\psi}\sum_{k=1}^{2}\frac{\mu_k z_k}{r_k^5}(y_k - jx_k) - p_y \tag{14}$$

Here we have used the fact that $p_x = 0$. For the next term in the expansion of Equation 10 we cannot include the complete linear expression in ζ because this would not permit the analytical solution that follows. Fortunately, however, the greatest contribution to this term is a pure rotation: the regression of the nodes by -4.9 deg/year caused by the Earth's lowest zonal term. We introduce the following real-valued function, which is positive and dominated by the constant zonal term.

$$h(t) = 1.5\,\psi\,J_2(\frac{R}{A})^2 + \frac{0.75}{\psi}\sum_{k=1}^{2}\frac{\mu_k}{r_k^5}(r_k^2 - 3z_k^2) + p_z \tag{15}$$

Above we have inserted $\mu = A^3\psi^2$ and retained elements of the Sun and Moon perturbations that contribute to a pure rotation, which are larger than the remainder. The approximate free-drift differential equation for the orbit pole is now, instead of Equation 10:

$$\frac{d\zeta}{dt} = \gamma(t) - j\,h(t)\,\zeta \tag{16}$$

Since the last term of Equation 16 is a pure rotation, i.e. $h(t)$ is restricted to be real, the right-hand side of Equation 16 is not the most general first-order expansion in ζ. As a consequence, the rest term Q of the truncation contains both a first- and a second-order term in ζ:

$$|Q| < C_1|\zeta| + C_2|\zeta|^2 \le C_1\,i_{max} + C_2\,i_{max}^2 \tag{17}$$

The following coefficients are obtained by straightforward numerical calculation of the remainder:

$$C_1 = 0.0042 \text{ year}^{-1} \qquad C_2 = 0.03 \text{ rad}^{-1} \text{ year}^{-1}$$

We see that C_1, which is the unmodelled part of the linear term in Equation 16, is about 4 % of the modelled part $h(t) \approx 0.1 \text{ year}^{-1}$.

The station keeping problem can now be formulated as follows: A sequence of manoeuvres $\Delta\zeta_m$ and manoeuvre times t_m, $m = 1,2,...M$ must be found such that the inclination vector stays inside the deadband

$$|\zeta(t)| \le i_{max} \tag{18}$$

during the mission lifetime $0 \le t \le T$ when exposed to the combined effect of the drift and the manoeuvres. Use of the station keeping fuel is optimised by minimising the expression

$$F = \sum_{m=1}^{M} |\Delta\zeta_m| \tag{19}$$

We set $t = 0$ at the mission start. The corresponding value of ζ is taken to be

$$\zeta(0) = \zeta_0 \qquad |\zeta_0| \le i_{max} \tag{20}$$

We can solve the free-drift Equation 16 explicitly by introducing $H(t)$, which is obtained by numerical integration of Equation 15,

$$H(t) = \int_0^t h(\tau)d\tau \tag{21}$$

By moving the last term of Equation 16 to the left-hand side and multiplying both sides by $\exp[j\,H(t)]$ we obtain

$$\frac{d}{dt}\{\zeta \exp[j\,H(t)]\} = \gamma(t)\,\exp[j\,H(t)] \tag{22}$$

We can now solve ζ after integrating both sides of Equation 22. By including also the effect of the inclination manoeuvres carried out before the current time t we obtain

$$\zeta \exp[j\,H(t)] = \zeta_0 + \Gamma(t) + \sum_{m:t_m < t} \Delta\zeta_m \exp[j\,H(t_m)] \tag{23}$$

where we obtain the complex-valued integral $\Gamma(t)$ by straight-forward numerical integration of Equation 14 using Equation 21

$$\Gamma(t) = \int_0^t \gamma(\tau)\,\exp[j\,H(\tau)]\,d\tau \tag{24}$$

For the following optimisation it will turn out to be an advantage to transform the dependent variable ζ to η via

$$\eta(t) = \zeta(t)\,\exp[j\,H(t)] \tag{25}$$

which has the same initial value as ζ

$$\eta(0) = \zeta_0 \tag{26}$$

The manoeuvres $\Delta\zeta_m$ are similarly transformed to

$$\Delta\eta_m = \Delta\zeta_m \exp[j\,H(t_m)] \tag{27}$$

At this point we see the advantage that $h(t)$ and $H(t)$ are real functions. The transformation from ζ to η is a pure rotation, so the station keeping (Equation 18) and fuel optimisation (Equation 19) criteria retain the same form:

$$|\eta(t)| \le i_{\max} \tag{28}$$

$$F = \sum_{m=1}^{M} |\Delta\eta_m| \tag{29}$$

The evolution of the transformed inclination vector $\eta(t)$ under the influence of the free drift and the inclination manoeuvres is now analogous to Equation 23:

$$\eta(t) = \zeta_0 + \Gamma(t) + \sum_{m:t_m < t} \Delta\eta_m \tag{30}$$

Figure A shows $\Gamma(t)$, the free-drift part of the transformed inclination vector η and, for comparison, the corresponding curve for the untransformed inclination vector ζ given by $\exp[-jH(t)]\Gamma(t)$. In order to optimise the manoeuvres $\Delta\eta_m$ we must first look at the solution to Equation 30 and the criterion Equation 28 at mission end $(t = T)$

$$|\eta(T)| = |\zeta_0 + \Gamma(T) + \sum_{m=1}^{M} \Delta\eta_m| \le i_{max} \tag{31}$$

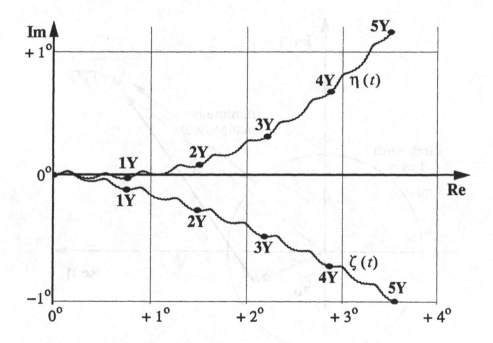

Figure 6.4.A. The drifts of the untransformed $\zeta(t)$ and the transformed $\eta(t)$ inclination vectors without manoeuvres during 5 years, starting from 0 on 1995 January 1. The scaling is converted from radians to degrees.

The simple geometric interpretation of the above equation is shown in Figure B. The series of manoeuvres must link the point $\zeta_0 + \Gamma(T)$ with the circle about 0 with radius i_{max}. The result of the optimisation is that the shortest path is a straight line from the point outside the circle, directed towards the centre of the circle, but stopping at the circumference. In general:

$$F = \sum_{m=1}^{M} |\Delta\eta_m| \geq |\zeta_0 + \Gamma(T)| - i_{max} \tag{32}$$

The minimum is attained when all $\Delta\eta_m$ have the same direction, namely opposite to $\zeta_0 + \Gamma(T)$:

$$\arg \Delta\eta_m = \pi + \arg[\zeta_0 + \Gamma(T)] \quad \text{for all } m = 1,2,...M \tag{33}$$

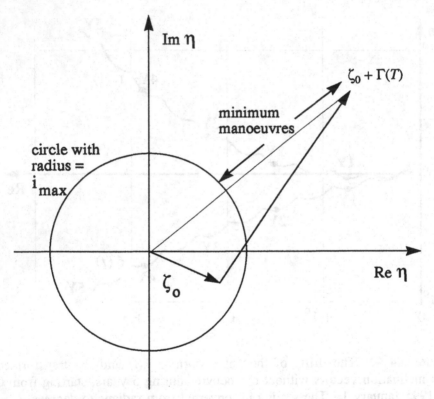

Figure 6.4.B. Optimum solution is the shortest path from a point outside a circle to the periphery of the circle.

We assume, of course, that the right-hand side of Equation 32 is positive, otherwise inclination station keeping would not be needed. The above results do not contain any indication of suitable manoeuvre times t_m. We will investigate whether we can select them such that the station keeping criterion

$$|\eta(t)| = |\zeta_0 + \Gamma(t) + \sum_{m:t_m < t} \Delta\eta_m| \leq i_{max} \tag{34}$$

is satisfied throughout the mission $0 \leq t \leq T$ using the optimal manoeuvre direction (Equation 33) for each $\Delta\eta_m$. For this purpose it is easier to visualise Equation 34 as the requirement to keep each point of the curve

$$\zeta_0 + \Gamma(t) \quad \text{for all} \quad 0 \leq t \leq T \tag{35}$$

inside a circle with radius i_{max} and the centre at

$$\rho(t) = -\sum_{m:t_m < t} \Delta\eta_m \tag{36}$$

Each new manoeuvre $\Delta\eta_m$ shifts the circle in the complex plane by $-\Delta\eta_m$ according to Equation 36. The curve of Equation 35 must be contained inside the envelope of the circles with radius i_{max} and centres at the points of Equation 36, as shown in Figure C. If this is possible when all manoeuvres have the optimal direction, one can perform optimal inclination manoeuvres during the whole mission.

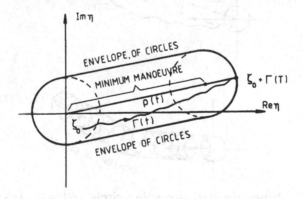

Figure 6.4.C. The optimal manoeuvre direction of Figure B can be used throughout the mission if the curve $\zeta_0 + \Gamma(t)$, $0 \leq t \leq T$ lies inside the envelope of circles with radii = i_{max}.

The fuel requirement is then obtained from the global minimum

$$F_{min} = |\rho(T)| = |\zeta_0 + \Gamma(T)| - i_{max} \tag{37}$$

The manoeuvres must be made such that the curve of Equation 35 always lies inside one of the circles, but there is nevertheless a certain margin for selecting the manoeuvre times t_m and the size of each manoeuvre $|\Delta\eta_m|$. The global optimum is still achieved as long as the direction of each manoeuvre $\Delta\eta_m$ coincides with the optimal direction of Equation 33. The untransformed manoeuvre $\Delta\zeta_m$ will, of course, then change in direction with time t_m according to Equation 27.

Following is the solution for the situation in which the curve of Equation 35 is not completely inside the envelope of the circles. This is likely to be the case if the station keeping deadband i_{max} is narrow or the mission lifetime is long. Figure D shows a case where the curve of Equation 35 lies partly outside the envelope of circles described above with the consequence that the optimal direction of $\Delta\eta_m$ cannot be used everywhere. Instead, one must construct another envelope of circles that covers the curve completely but where the centres $\rho(t)$ of the circles no longer lie on a straight line. The resulting curve described by $\rho(t)$ must be made as short as possible because its length equals the fuel penalty function F.

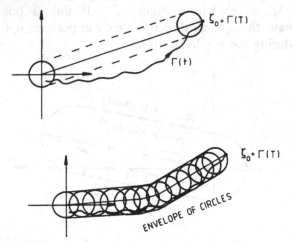

Figure 6.4.D. When the station keeping circle $|\eta| = i_{max}$ is small, one may have to manoeuvre along a curved line in the η-plane (lower figure). The envelope of circles no longer follows a straight line.

One can see from Figure D that the optimal solution in this case is an envelope of circles where the boundary follows and touches the curve $\zeta_0 + \Gamma(T)$ at its most protruding bend but otherwise tries to follow a straight line. Before quantifying this statement, we must introduce a small margin d to be inserted between the curve $\zeta_0 + \Gamma(T)$ and the boundary of the envelope of circles. Without this margin, we would need to perform station keeping manoeuvres too frequently at the times when the curve touches the boundary. We now introduce the two constraining functions

$$B_1(t) = \max_{0 \leq \tau \leq t} \left[\arg(\zeta_0 + \Gamma(\tau)) - \arcsin \frac{i_{max} - d}{|\zeta_0 + \Gamma(\tau)|} \right] \tag{38}$$

$$B_2(t) = \min_{0 \leq \tau \leq t} \left[\arg(\zeta_0 + \Gamma(\tau)) + \arcsin \frac{i_{max} - d}{|\zeta_0 + \Gamma(\tau)|} \right]$$

It is now easy to see, from Figure E, that the optimal solution for the first manoeuvre $\Delta\eta_1$ is a direction as close as possible to the optimal direction of Equation 33, i.e.

$$\arg \Delta\eta_1 = \pi + \arg[\zeta_0 + \Gamma(T)] \tag{39}$$

but which satisfies the constraints

$$B_1(T) \leq \arg \Delta\eta_1 - \pi \leq B_2(T) \tag{40}$$

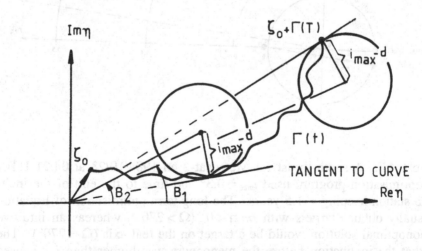

Figure 6.4.E. Example of the angles $B_1(T)$ and $B_2(T)$. The constraint $\arg \Delta\eta_1 - \pi \leq B_2(T)$ overrides the optimal direction $\arg[\zeta_0 + \Gamma(T)]$

It may happen for long missions that the lower constraint in Equation 40 becomes greater than the upper constraint. If there should be a value of $t = t_N$ for which

$$B_1(t) \geq B_2(t) \text{ when } t \geq t_N \tag{41}$$

then the time T in the inequality of Equation 40 must be replaced by t_N and the solution reduces to

$$\arg \Delta\eta_1 - \pi = B_1(t_N) = B_2(t_N) \tag{42}$$

Since the manoeuvres $\Delta\eta_m$, $m = 1,...M$ no longer have the same direction, it is easiest to recalculate the constrained optimisation, given above, anew for each manoeuvre. One takes the initial value ζ_0 = the inclination vector just before the manoeuvre and T = the time from the present to the end of the mission.

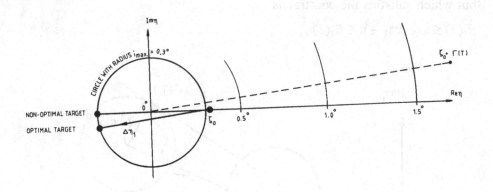

Figure 6.4.F. A south thrust for Meteosat-2 on 1984/10/23 at 04:21 UTC. The optimisation program used $i_{max} = 0.3°$ and time to the end of the inclination station keeping = 1.5 years. The long-term optimisation of Equation 33 usually obtains targets with $Im\,\eta < 0$ ($\Omega > 270°$), whereas an intuitive but nonoptimal solution would be a target on the real axis ($\Omega = 270°$). The fact that the inclination before the manoeuvre was higher than i_{max} caused no problem in this case.

6.5 Influence of Errors

Ideally, an inclination thrust should change only the out-of-plane component of the spacecraft's orbital velocity, but not its absolute size or the in-plane component. To this end the direction of the thrust velocity increment, $\Delta \overline{V}$, should be orthogonal to the plane bisecting the planes of the two orbits, before and after the manoeuvre, Figure A.

If $\Delta \overline{V}$ deviates by a small angle d from the ideal direction, one obtains an in-plane component $\Delta V \sin d$ of the thrust that changes the longitude drift rate and the eccentricity. Since inclination thrusts are more than 20 times greater than longitude thrusts, as can be seen in Tables 2 and 3, even small deviation angles of $2°$ or $3°$ cause in-plane disturbances of the same size as what is used for the longitude station keeping. For the same reason one can neglect the converse effect, since the influence of the out-of-plane component of longitude manoeuvres on the inclination is very small.

The expected thrust direction error angle d is different for different types of spacecraft. Usually spin-stabilised spacecraft are stabilised to better than one degree, even during thruster firing. It is, however, not always necessary to adjust the spacecraft attitude before a small inclination manoeuvre, so a small but predictable in-plane component may be obtained. Three-axis-stabilised spacecraft with an on-board control system are usually aligned relative to the plane of the orbit. The inclination manoeuvre thrust is then aligned orthogonal to the plane of the orbit as it was before the manoeuvre. It may happen, however, that the firing of the thrusters disturbs the control, in particular if the thruster plumes impinge on the solar array panels. As a result, one may obtain a thrust tip-off of several degrees with a correspondingly large in-plane $\Delta \overline{V}$ component.

If the size and direction of the deviation can be predicted, one can make use of the in-plane component for the longitude control. If one cannot predict the direction of the deviation, it is advisable to perform inclination manoeuvres when the spacecraft is as far as possible near the centre of the longitude deadband. In this way one has the best possibility to determine the longitude drift change after the manoeuvre and perform a corrective longitude thrust without violating the longitude deadband. These corrective thrusts then increase the fuel budget needed for the inclination station keeping.

Errors in the thrust size influence the resulting $\bar{i}$ and its subsequent drift, by a parallel shift to the right or left along the nominal $\Delta\bar{i}$-direction. If it is important for the mission that no overshoot occurs, the Operator must subtract a sufficiently wide margin from the planned manoeuvre ΔV. If one obtains an undershoot of the thrust, the subsequent free drift period becomes shorter than planned, since $\bar{i}$ reaches the right-hand boundary sooner than desired. It is usually not worth while to perform a correction thrust immediately in this case. Instead the correction is incorporated into the next scheduled station keeping manoeuvre. Over- or undershoot of the thrusts does not affect the total ΔV-budget for station keeping, when it is expressed in m/s. It only influences the fuel budget, in kg, because the detection of an over- or undershoot makes it necessary to recalculate the conversion rate from m/s to kg of fuel.

Orbit determination errors do not influence the accuracy of the direction of $\Delta\bar{i}$ by any appreciable amount. The accuracy in the determination of the spacecraft longitude is usually better than fractions of degrees. The direction of $\Delta\bar{i}$ depends on the knowledge of the sidereal angle of the spacecraft at the thrust, which is derived from the longitude. An error in the direction of even one degree would not be of significance for the manoeuvre. Of more importance is the influence of the orbit determination errors on the knowledge of $\bar{i}$. This can be visualised by replacing the point $\bar{i}$ in a plot of the inclination vector plane by a small error circle. The radius of this circle stays constant in time and is proportional to the error in the out-of-plane components of the orbit determination. The circle moves in the same way as already described for the point $\bar{i}$ under the influence of the natural drift and the inclination manoeuvres.

If the requirement on the station keeping is that the inclination shall stay below i_{max} with a probability corresponding to a 3σ error in the orbit determination, one must make sure that the 3σ error circle around $\bar{i}$ stays inside the circle with radius i_{max}. This is done according to the usual station keeping strategies, after subtracting the 3σ error margin from i_{max}. Other requirements, such as a 2σ probability level etc., are handled in an analogous way.

The inclination station keeping fuel requirements are only weakly affected by the orbit determination error. The slight increase in fuel is only due to the fact that i_{max} is decreased by the subtraction of the inclination error margin.

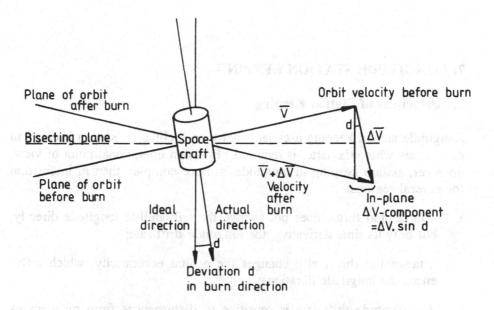

Figure 6.5.A. If the inclination thrust direction $\Delta\overline{V}$ deviates by an angle d from the direction orthogonal to the plane bisecting the orbit planes, the magnitude of the orbital velocity is changed by the in-plane component $\Delta V \sin d$.

7. LONGITUDE STATION KEEPING

7.1 Principles of Station Keeping

Longitude station keeping uses less fuel than inclination station keeping in most cases where the latter is required. From an operational point of view, however, station keeping in longitude is more complex than in inclination for several reasons:

- a tangential thrust does not change the subsatellite longitude directly, but only its time derivative, the longitude drift rate;

- a tangential thrust also changes the orbital eccentricity, which influences the longitude librations;

- the longitude drift rate is sensitive to disturbances from manoeuvres of attitude (Section 5.1) and inclination (Section 6.5).

The longitude deadband is usually expressed in the form of its midpoint λ_m and its half-width $\delta\lambda$ as $\lambda_m \pm \delta\lambda$, which means that λ shall lie in the interval

$$\lambda_m - \delta\lambda \leq \lambda \leq \lambda_m + \delta\lambda$$

Typical values of $\delta\lambda$ range from a few times $0.01°$ up to $1°$, but seldom lower or higher. A very small deadband is difficult to maintain since it has to accommodate both the short term variations in the longitude and the effect of non-zero values of the longitude drift, the eccentricity and the inclination. We have seen in Section 4.3 that the Sun and Moon cause longitude variations of up to $0.0041°$. In most cases the longitude drift and the librations from the eccentricity dominate, whereas the inclination contributes only with the second order effect ("Figure-of-eight") of Section 2.3.

For a conservative station keeping one subtracts a margin from $\delta\lambda$ according to Section 5.1 in order to ensure that the deadband is not violated even under the influence of errors. The errors from an orbit determination at time t_0 affect the accuracy of the predicted longitude according to a formula in Section 2.3

$$\varepsilon_\lambda = \varepsilon_{\lambda 0} + \varepsilon_D \psi(t - t_0) + 2\varepsilon_e$$

Here ε with an index denotes the error of the parameter in question. We see that one term increases linearly with the time elapsed since the latest orbit determination. This comes from the error in the longitude drift rate, which is proportional to the error in the semimajor axis. For this reason it is advisable to plan longitude manoeuvres shortly before their execution, preceded by an extra orbit determination.

The error $\delta\Delta V$ of a planned tangential thrust ΔV causes an error in the longitude of the predicted orbit of

$$|(\delta\Delta V/V)[4\sin(s - s_b) - 3(s - s_b)]|$$

It can be compensated for either by subtracting a margin from the thrust ΔV or from the longitude deadband $\delta\lambda$. One can also execute the manoeuvre without margin, determine the orbit and perform a correction thrust later if it turns out that the obtained error would cause the deadband to be violated.

Errors in orbit determination or manoeuvre execution are of importance only through the size of ΔV. On the other hand, the resulting orbit is quite insensitive to errors in the direction of $\Delta \overline{V}$, in the time of thrust and in the spacecraft sidereal angle at the thrust. These errors are normally neglected for all practical calculations.

The level of sophistication of the station keeping preparations depends on the deadband size, the tangential acceleration due to the Earth's asymmetry at the station keeping longitude and the size of the solar radiation perturbation. We will in this and the following two sections deal with station keeping away from any of the four equilibrium longitudes and treat the near-equilibrium station keeping in Section 7.4.

The tangential acceleration on the spacecraft that is due to the tesseral terms in the Earth's gravity field, B, is described in Section 4.2 and listed in Table 2. It causes an acceleration of the mean spacecraft longitude

$$\ddot{\lambda} = -3B/A$$

This is listed in the third column of Table 2, converted to degrees/day^2 through multiplication by $(180/\pi) \times (24 \times 60 \times 60)^2$.

To start with we will now consider a simple type of station keeping with a relatively high $\delta\lambda$, typically $1°$. The longitude acceleration in Table 2 can be considered to be constant in the interval $\lambda_m \pm \delta\lambda$. The mean longitude, disregarding short-term variations and librations due to the eccentricity, describes a parabola as a function of time t

$$\lambda(t) = \lambda_0 + \dot{\lambda}_0 (t - t_0) + 0.5\, \ddot{\lambda}\, (t - t_0)^2$$

The sign of $\ddot{\lambda}$ decides which way the parabola is turned, Figure A. The following formulas are valid when it is positive, whereas a negative value is dealt with in an analogous way by swapping east and west and changing the signs. The maximum length T of the station keeping cycle is obtained by selecting the two constants λ_0 and $\dot{\lambda}_0$ of the parabola such that

$\lambda(t_0) = \lambda_m + \delta\lambda$ start cycle with longitude at east boundary
$\dot{\lambda}(t_0) = \dot{\lambda}_0$ start cycle with optimal longitude drift rate
$\lambda(t_0 + T/2) = \lambda_m - \delta\lambda$ touch west boundary
$\dot{\lambda}(t_0 + T/2) = 0$ natural longitude drift reversal at west boundary

This gives the solution

$$\lambda_0 = \lambda_m + \delta\lambda \quad ; \quad \dot{\lambda}_0 = -2\sqrt{\ddot{\lambda}\,\delta\lambda} \quad ; \quad T = 4\sqrt{\delta\lambda / \ddot{\lambda}}$$

Note that $\delta\lambda$ denotes half the size of the deadband $\lambda_m \pm \delta\lambda$. One can calculate the cycle length T in days by inserting above $\delta\lambda$ in degrees and $\ddot{\lambda}$ in degrees/day^2 from Table 2. At the end of the cycle we are back at the east boundary with a drift rate that has the same size as the starting drift but the opposite sign:

$$\lambda(t_0 + T) = \lambda_m + \delta\lambda \quad ; \quad \dot{\lambda}(t_0 + T) = -\dot{\lambda}_0$$

At this place we can start a new station keeping cycle with (ideally) an identical free drift parabola by changing the mean longitude drift rate by

$$\Delta\dot{\lambda} = \dot{\lambda}(t_0) - \dot{\lambda}(t_0 + T) = 2\dot{\lambda}_0 = -4\sqrt{\ddot{\lambda}\,\delta\lambda}$$

This change in drift rate is, according to Section 3.3, obtained by performing an east thrust of the size

$$\Delta V = -A\Delta\dot{\lambda}/3 = (4A/3)\sqrt{\ddot{\lambda}\,\delta\lambda}$$

With the same station keeping thrust executed every T days one obtains at the average, over several cycles, the following change in the mean longitude drift rate. This equation is valid for both positive and negative longitude accelerations.

$$\Delta\dot{\lambda}/T = -\ddot{\lambda}$$

Under the same conditions, the corresponding average of the executed ΔV becomes the same as the natural acceleration B but with opposite sign:

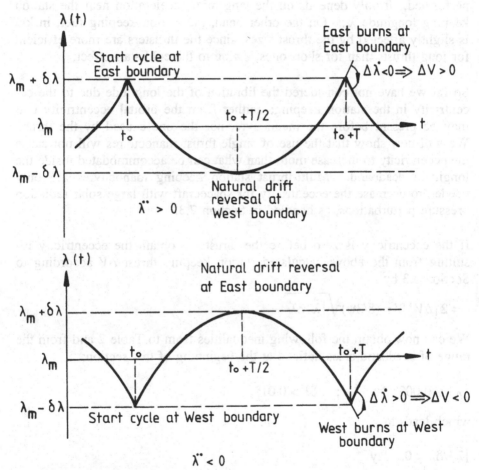

Figure 7.1.A. The parabola described by the mean longitude (λ) during the ideal station keeping cycle is shown at the top for a positive longitude acceleration and at the bottom for a negative.

$$\Delta V/T = A\ddot{\lambda}/3 = -B$$

This result should be expected, since the manoeuvres are performed to compensate for B. One can now directly calculate the mean ΔV in m/s per year for longitude station keeping by multiplying $|B|$ by the number of seconds in a year. This is listed in the right-hand column of Table 2.

It is important to note that the mean station keeping ΔV is independent of the deadband width $2\delta\lambda$ and then also of how often the manoeuvres are performed. It only depends on the tangential acceleration near the station keeping longitude λ_m. On the other hand, the station keeping fuel, in kg, is slightly affected by the thrust sizes, since the thrusters are more efficient for long thrusts than for short ones, owing to the cold-start effect.

So far we have not considered the libration of the longitude due to the eccentricity in the station keeping; neither from the orbital eccentricity that may be present before the manoeuvre, nor the one caused by the thrust. We will now show that the use of single thrust manoeuvres will not cause the eccentricity to increase more than what can be accommodated inside the longitude deadband. Multi-thrust station keeping manoeuvres are only needed to decrease the eccentricity for spacecraft with large solar radiation pressure perturbations, as handled in Section 7.3.

If the eccentricity is zero before the thrust we obtain the eccentricity resulting from the above calculated station keeping thrust ΔV according to Section 3.3 by

$$e = 2|\Delta V|/V = (8/3\psi)\sqrt{|\ddot{\lambda}|\delta\lambda}$$

We can now obtain the following inequalities from to Table 2 and from the range of deadband sizes defined at the beginning of this section:

$$|\ddot{\lambda}| \le 0.002° \, \text{day}^{-2} \quad ; \quad \delta\lambda > 0.01°$$

which leads to

$$|\ddot{\lambda}|/\delta\lambda < 0.2 \, \text{day}^{-2}$$

The amplitude of the longitude librations caused by the eccentricity becomes $= 2e$, according to Section 2.3. We can see from the inequality above that

$$2e = (16/3\psi)\sqrt{|\ddot{\lambda}|\,\delta\lambda} < (16/3\psi)\sqrt{0.2}\;\delta\lambda = 0.38\,\delta\lambda$$

In order to obtain consistency with the units one must insert here the value $\psi = 6.30$ rad/day. The result means that the librations $= 2 \times$ amplitude will never fill more than about one third of the deadband $= 2\delta\lambda$, which leaves at least two thirds of it for the station keeping cycle of the mean longitude drift.

For the case when the eccentricity is not zero before the thrust one selects the timing of an east thrust to be near the orbit apogee in order to decrease the eccentricity as much as possible. Before the thrust the spacecraft sidereal angle at apogee is in the direction in which $-\bar{e}_{old}$ is pointing. The resulting $\Delta\bar{e}$ is antiparallel to $\bar{e}_{old}$, so $|\bar{e}_{new}|$ must be smaller than at least one of $|\bar{e}_{old}|$ or $|\Delta\bar{e}|$. In the same way, a west thrust is executed near the orbit perigee, at the sidereal angle of the $+\bar{e}_{old}$ direction.

By performing longitude station keeping manoeuvres in this way near apogee (or perigee), one can prevent the eccentricity from becoming too large from the single thrusts. Often the time of the manoeuvre is not very critical, but one may execute the thrusts within the same semicircle of the orbit, i.e. within ± 6 hours from the apogee or perigee. More consideration must be given to the eccentricity vector only when it is changed considerably by the solar radiation pressure, Section 7.3. Section 7.2 gives more details about the longitude control in practical station keeping.

7.2 Longitude Control

The theoretical model for station keeping, described at the end of the previous section, can be used, after some modifications, for practical operations. Because of the propagation of errors, one cannot expect to use exactly the same size of ΔV and T for each cycle. The station keeping control cycle is stable only if each ΔV is calculated from the actually determined and predicted orbit, based upon as recent tracking data as possible. Still, the mean fuel consumption, averaged over several cycles, remains the same as that given in Table 2.

One can use the prints of the predicted orbit shown in Figures 5.2.B and 5.2.C to calculate the time and size of the station keeping ΔV in a rough manner in the following way. From Figure 5.2.C we see that the maximum or minimum longitude will pass beyond one of the longitude boundaries on a certain day. From Figure 5.2.B we select a suitable time t_1 before this day for the manoeuvre. If the longitude is near the east boundary one prepares an east thrust at the apogee, where $v \, (= TR.AN) = 180°$. In the same way, at the west boundary one makes a west thrust at the perigee, where $v = 0°$.

We read off the mean longitude λ_1 and its drift rate $\dot{\lambda}_1$ (deg/day) at the day of the thrust from Figure 5.2.C and $\ddot{\lambda}$ (deg/day^2) from Table 2. The thrust size is calculated (in m/s) as the sum of a term that removes $\dot{\lambda}_1$ and another term that starts a drift with a natural reversal at the opposite boundary:

$$\Delta V = 2.84 \left(\dot{\lambda}_1 + \sqrt{2\ddot{\lambda} \, (\lambda_1 - \lambda_m + \delta\lambda)} \; \right) > 0$$

when $\ddot{\lambda} > 0$; $\dot{\lambda}_1 > 0$; $\lambda_1 > \lambda_m - \delta\lambda$

$$\Delta V = 2.84 \left(\dot{\lambda}_1 - \sqrt{2\ddot{\lambda} \, (\lambda_1 - \lambda_m - \delta\lambda)} \; \right) < 0$$

when $\ddot{\lambda} < 0$; $\dot{\lambda}_1 < 0$; $\lambda_1 < \lambda_m + \delta\lambda$

This rough calculation of ΔV only considers the evolution of the mean longitude and disregards the libration in longitude around the mean value due to the eccentricity and the natural short-term variations. In order to take these into consideration at the ΔV preparation, one can follow one of two possible methods:

•   In the semianalytic method, one subtracts from the deadband, once for the whole mission, longitude margins for eccentricity librations and short-term variation, etc. In what remains of the deadband one fits the parabola of the mean longitude free drift, Figure A. One must make sure that the eccentricity always stays below the limit that corresponds to the libration margin.

•   A numerical method can be used to calculate a ΔV such that (Figure B) the actually predicted longitude touches the opposite side of the deadband. Margins are only needed for errors in orbit determination and thrust performance prediction.

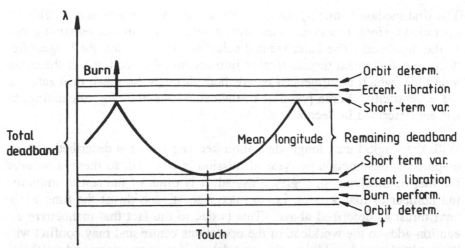

Figure 7.2.A. Semianalytic station keeping thrust preparation by fitting the free drift motion of the mean longitude into what remains of the deadband after subtraction of margins for libration etc.

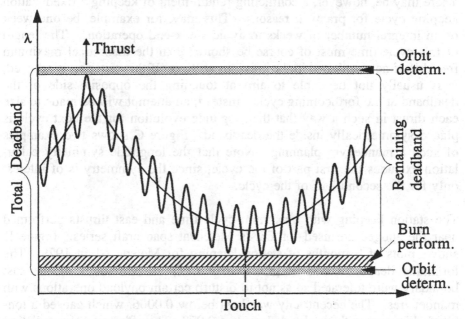

Figure 7.2.B. Station keeping preparation by numerical integration of the actual longitude. Margins are needed only for orbit determination and thrust performance prediction errors.

The first method is mainly used for mission analysis, but is less suitable for operations, since it leaves unnecessarily little space in the remaining part of the deadband. The latter method calculates in each case the longest free drift time to the next thrust. It takes into account the evolution of the actual longitude, which is influenced by ΔV both through the mean drift rate and the eccentricity. Some practical mathematical methods for calculating the ΔV are described in Section 7.6.

With the single thrust longitude station keeping we have described, the average fuel consumption per year of mission is constant, so there is no need for fuel optimisation strategies. Instead, it is often of interest to minimise the number of manoeuvres, i.e. to maximise at each thrust the time to the next thrust as described above. This is due to the fact that manoeuvre execution adds to the workload of the operations centre and may conflict with the payload use. In addition, the orbit determination accuracy and reliability are degraded by too frequent orbit manoeuvres.

There may be, however, a conflicting requirement of keeping a fixed station keeping cycle for practical reasons. This may, for example, be one week or an integral number of weeks to avoid week-end operations. The length of this cycle time must of course be shorter than the theoretical maximum free drift time mentioned in the previous section. If this criterion is adopted, it is usually not desirable to aim at touching the opposite side of the deadband at the forthcoming cycle. Instead, an attempt will be made to size each thrust in such a way that the longitude evolution for the next cycle is placed symmetrically inside the deadband. Figure C shows two examples of such a manoeuvre planning. Note that the longitude symmetry calculation excludes the first part of the cycle, since the symmetry is of interest only for the second half of the cycle.

The station keeping with maximum cycle time and east thrusts performed near the apogee are used for ESA's Meteosat spacecraft series. Figure D shows plots of 9 months of station keeping for Meteosat-1 in 1979. The longitude deadband was in principle $0° \pm 1°$, but violations of the east boundary were tolerated so as not to disturb certain payload operations with manoeuvres. The eccentricity was kept below 0.0006, which caused a longitude libration with amplitude less than $0.07°$. The effective solar radiation cross-section to mass ratio was 0.015 m^2/kg.

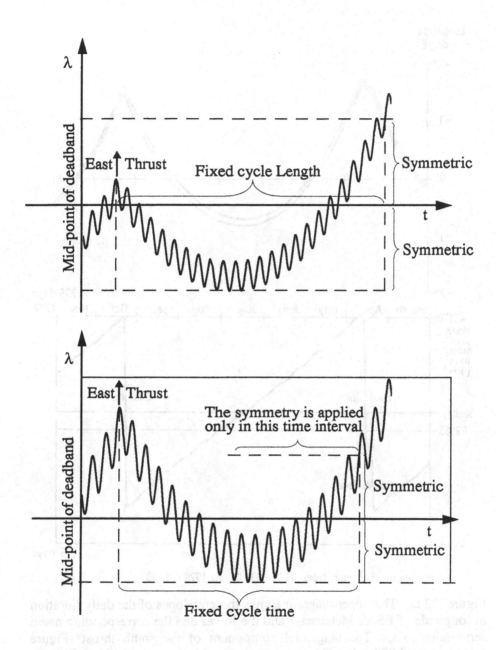

Figure 7.2.C. Station keeping with fixed cycle time, where the longitude at the later half of the cycle is placed symmetrically inside the deadband.

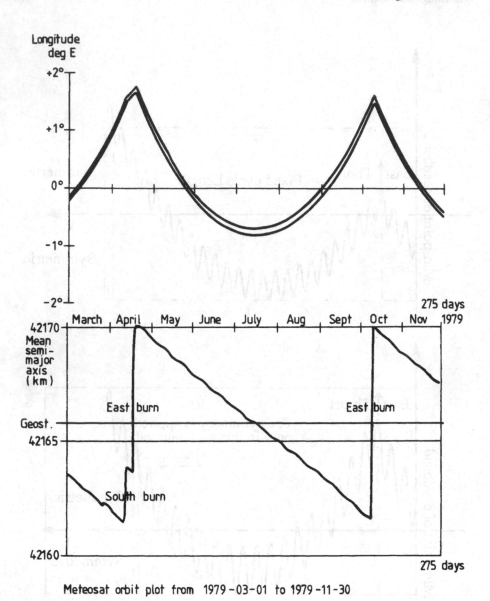

Meteosat orbit plot from 1979−03−01 to 1979−11−30

Figure 7.2.D. The upper diagram shows the envelopes of the daily libration in longitude of ESA's Meteosat-1 and the lower one the corresponding mean semimajor axis. The tangential component of the south thrust (Figure 6.1.A) was +0.073 m/s. The two east thrusts were, respectively, 0.247 m/s (2% undershoot) and 0.319 m/s (0.3% undershoot).

With this method of selecting the time of the thrust one cannot exclude the possibility that it coincides with an eclipse, see Section 5.3. If the spacecraft is built in such a way that one cannot perform manoeuvres during or near an eclipse, one has to change the thrust time to the nearest allowed possibility, before or after the eclipse. The thrust at the new execution time must of course be tangential to the orbit at the mid-point of the actual execution. The size of the new ΔV is slightly different, since its effect on the eccentricity vector depends on the time of execution.

The longitude of the free drift parabola apsis, where the natural drift reversal takes place, is very sensitive to small errors in the thrust size. In spite of the longitude margin for thrust performance, the apsis may still happen to violate the deadband if the thrust overperforms more than expected. This can be prevented by the execution of a drift slow-down corrective thrust before violation of the deadband, Figure E.

It now remains to determine under which conditions the simple eccentricity control, with thrusts near apogee or perigee, can be applied. We saw at the end of the preceding section that each thrust either decreases the eccentricity or increases it temporarily by a small amount. We must then check if the eccentricity decrease is sufficiently great to eliminate the increase caused by the solar-radiation pressure.

From Section 3.3 we get the change of the eccentricity vector from a tangential ΔV each cycle T:

$$\frac{|\Delta \bar{e}|}{T} = \frac{2}{V} \frac{|\Delta V|}{T} = \frac{2}{V} |B|$$

This must keep up with the change in $\bar{e}$ from the solar radiation pressure from Section 4.5:

$$\left| \frac{\delta \bar{e}}{\delta t} \right| = \frac{3P\sigma}{2V}$$

The condition becomes

$$\frac{3P\sigma}{2V} \le \frac{2}{V} |B| \quad \text{or} \quad \sigma \le \frac{4|B|}{3P}$$

The highest value of $|B|$ is $= 0.652 \times 10^{-7}$ m/s², which yields $\sigma \le 0.019$ m²/kg. For such small values of σ the drift of $\bar{e}$ is dominated by the intermediate-term lunar-gravity perturbation and not by the solar radiation pressure, according to Section 4.5.

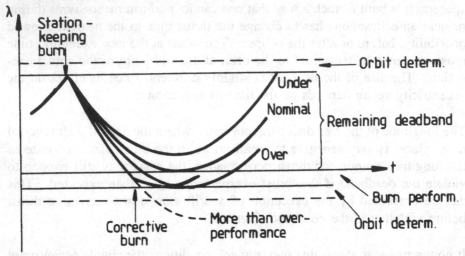

Figure 7.2.E. If the station keeping thrust overperforms more than planned, one can slow down the drift by an extra corrective thrust in the opposite direction. The librations from the longitude are omitted here to make the diagram more easily readable.

7.3 Eccentricity Control

Three-axis stabilised spacecraft with high demand on onboard electric generation power often have extending solar panels with a large surface. When the spacecraft has a high cross-section to mass ratio, the solar radiation pressure will cause the orbit eccentricity to increase more than can be reduced in the way described in the preceding section. Since this is often combined with a narrow longitude deadband, particular care has to be taken with the eccentricity control during station keeping.

A very high solar radiation acceleration can be compensated for only by multi-thrust station keeping, as described at the end of this section. Before this, however, we will show an optimal single thrust method that can be used for spacecraft with intermediately high cross-section to mass ratios. It is called the *sun-pointing-perigee* strategy.

The sun-pointing-perigee strategy consists in choosing a more optimal time for the longitude drift correction thrust than apogee or perigee. It cannot, however, prevent the eccentricity from remaining at a relatively high value, so the resulting longitude librations fill a considerable part of the deadband. The result is a short station keeping cycle.

The change in the eccentricity vector (Section 3.3) from each station keeping thrust

$$\Delta \bar{e} = \frac{2}{V} \Delta V \begin{pmatrix} \cos s_b \\ \sin s_b \end{pmatrix}$$

can, at the average, be approximated by a continuous change, where we insert $\Delta V/T = -B$ (B is the tangential acceleration from the Earth's gravity asymmetry of Table 2)

$$\frac{\Delta \bar{e}}{T} \to \frac{\delta \bar{e}}{\delta t} = -\frac{2}{V} B \begin{pmatrix} \cos s_b \\ \sin s_b \end{pmatrix}$$

Combining this with the change (Section 4.5) due to the solar radiation cross-section, we obtain, disregarding the intermediate-term variations from Section 4.3,

$$\frac{\delta \bar{e}}{\delta t} = \frac{3 P \sigma}{2V} \begin{pmatrix} -\sin s_S \\ \cos s_S \end{pmatrix} - \frac{2}{V} B \begin{pmatrix} \cos s_b \\ \sin s_b \end{pmatrix}$$

The task is to select all the s_b and the initial value of $\bar{e}$ such that $|\bar{e}|$ stays below as small a limit as possible during the whole mission. One can obtain the optimal solution

$$\bar{e}(t) = \left(\frac{3 P \sigma}{2V} - \frac{2}{V} |B| \right) \frac{Y}{2\pi} \begin{pmatrix} \cos s_S \\ \sin s_S \end{pmatrix}$$

by selecting $s_b = s_S + \pi/2$ when $B > 0$ and $s_b = s_S - \pi/2$ when $B < 0$. Here Y is the lenght of one year, expressed in seconds. With this solution the eccentricity vector rotates in a circle during one year, pointing towards the Sun with the sidereal angle $= s_S$ and has the constant size

$$|\bar{e}| = e = (Y/V\pi) (0.75 P\sigma - |B|)$$

This solution is analogous to the free-drift solution of $\bar{e}$ at the end of Section 4.5, but the radius of the eccentricity circle is reduced by an optimal use of the drift correction thrusts with the average magnitude $= |B|$. When $B > 0$, i.e. the longitude acceleration is negative, the station keeping west thrusts will be performed when the spacecraft is at the sidereal angle $s_b = s_s + \pi/2$, i.e. at local 18 hours. Similarly, when $B < 0$ and the longitude acceleration is positive the east thrusts are made at local 6 hours where $s_b = s_s - \pi/2$. For that reason, there is no conflict with eclipse constraint in this station keeping mode.

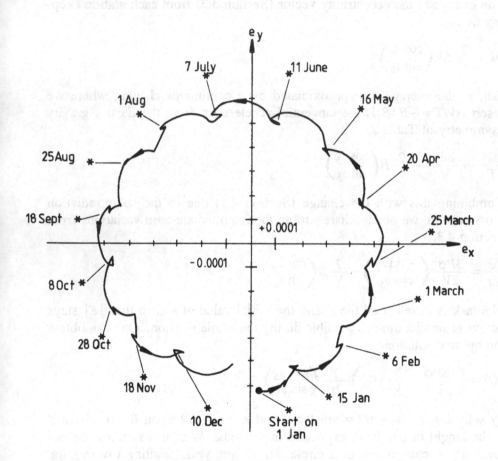

Figure 7.3.A. Eccentricity vector of one year of simulated station keeping with the sun-pointing-perigee strategy. The date of each thrust and the corresponding sun position are marked in the figure. The effective cross-section to mass ratio is 0.058 m²/kg giving a mean eccentricity of 0.00051. The longitude is shown in Figure B.

In reality $\bar{e}(t)$ will not move in a perfect circle, since the intermediate-term lunar gravity loops are superimposed and the manoeuvre thrusts are not executed continuously. Figures A and B show the results of a numerical integration of simulated station keeping for one year with 15 longitude manoeuvres. The curve described by $\bar{e}(t)$ still resembles a circle, since this method is used for spacecraft where the solar radiation pressure dominates over the intermediate-term lunar loops mentioned in Section 4.3.

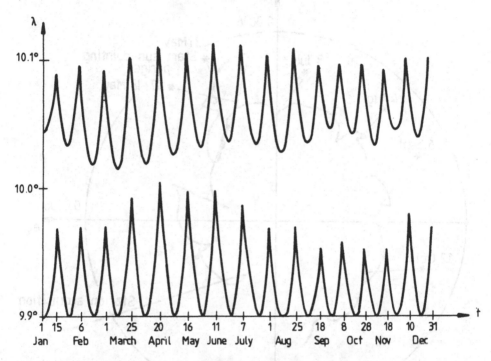

Figure 7.3.B. The envelope of the daily libration in longitude for the station keeping from Figure A fills about 0.12° of the 0.20° deadband width.

It now remains to describe how to operationally carry out station keeping according to the sun-pointing-perigee strategy. The local spacecraft time for the thrust given above is only approximately valid and depends on the actual orbit just before the manoeuvre. The optimal manoeuvre time has to be calculated by means of a computer program that aims at pointing the eccentricity vector, i.e. the perigee, to the Sun in the forthcoming station keeping cycle. Theoretically, the directions should coincide at the middle of the cycle, but in practice it is sufficient for them to coincide at some instant during the cycle. Figures C and D show orbit plots for the first 11 months of ESA's Marecs-A spacecraft, where the sun-pointing-perigee station keeping was applied for 6 months without inclination manoeuvres.

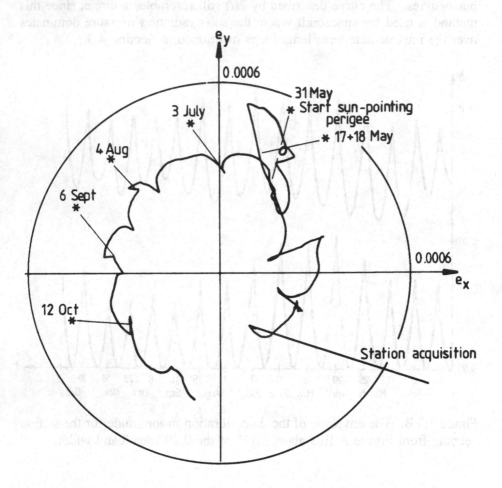

Figure 7.3.C. Eccentricity vector of ESA's Marecs-A spacecraft from station acquisition on 1982 January 2 until end of November. The sun-pointing-perigee strategy was started by a west-east thrust pair on May 17 and 18. The dates of the subsequent thrusts and of the corresponding Sun directions are marked in the figure. The effective cross-section to mass ratio is 0.036 m²/kg, giving a mean eccentricity of 0.00032. The longitude is shown in Figure D.

Section 7.6 describes some mathematical methods for calculating the station keeping manoeuvres. The target mean eccentricity to be used is obtained from the last formula. If we express the effective cross-section to mass ratio of the spacecraft σ in m²/kg and the acceleration of the asymmetric gravity field of the Earth at the station keeping longitude from Table 2, B in m/s² we can rewrite it in the form below. Here the right hand side must of course be positive. If it should be negative because σ is small, one shall instead apply the simpler station keeping mode described in the previous section.

$$e = (Y/V\pi)\,(0.75\,P\sigma - |B|) = 3267 \times (3.42 \times 10^{-6} \times \sigma - |B|)$$

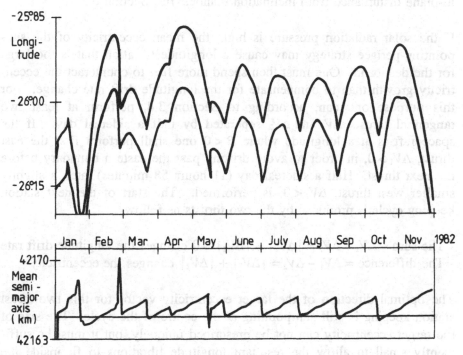

Figure 7.3.D. The envelope of the daily libration in longitude and the mean semi-major axis of the orbit of Marecs-A from Figure C. Each jump of the latter is due to a longitude station keeping manoeuvre. The longitude deadband was − 26° ± 0.15°. Plotted is the result of the orbit determination, except for the last month, which shows the predicted orbit. A station keeping manoeuvre is due at the end of November, but it was not yet planned at the time of the plot.

In order to start the sun-pointing-perigee strategy at the beginning of a mission one usually has to perform a station acquisition east-west manoeuvre in order to set up the eccentricity vector with the right size pointing to the Sun. However, if the initial eccentricity should happen to be sufficiently small one can wait with this acquisition until later when the eccentricity has grown such that the longitude librations become too large for convenient longitude station keeping. Such a reacquisition by an east-west thrust pair has to be performed in any case regularly since it is difficult in practice to maintain the sun-pointing-perigee manoeuvre strategy with only single thrusts for a long time because of accumulation of errors and in-plane disturbance from inclination manoeuvres, Section 6.5.

If the solar radiation pressure is high, the mean eccentricity of the sun-pointing-perigee strategy may cause a longitude libration that is too large for the deadband. One must then spend more fuel to counteract the eccentricity growth than to compensate for the longitude drift rate change. For this purpose one can, according to Section 3.4, perform at least two tangential thrusts ΔV_1 and ΔV_2 separated by half a sidereal day. If the spacecraft is at a longitude where $B < 0$ one shall perform first the east thrust, $\Delta V_1 > 0$, in order to avoid drifting past the eastern boundary before the next thrust. Half a sidereal day (11 hours 58 minutes) later, a slightly smaller west thrust, $\Delta V_2 < 0$, is performed. The start of the next station keeping cycle is produced by the two thrusts as follows:

- The sum $= \Delta V_1 + \Delta V_2 = |\Delta V_1| - |\Delta V_2| > 0$ changes the longitude drift rate
- The difference $= \Delta V_1 - \Delta V_2 = |\Delta V_1| + |\Delta V_2|$ changes the eccentricity

The optimal direction of the target eccentricity vector for this two-thrust station keeping is still sun-pointing at the centre of the cycle. The size of the target eccentricity can not be prescribed uniquely, but it must be sufficiently small to allow the resultant longitude librations to fit inside the longitude deadband and leave sufficient room for the parabola of the mean longitude drift. The smaller the eccentricity is made, the longer is the station keeping cycle, but the fuel increases. At the extreme, one obtains the maximum cycle by targeting for eccentricity zero at the centre of the next cycle, Figure E. The corresponding fuel consumption is

$$= |\Delta V| / T = 108 \times \sigma \text{ m/s/year, with } \sigma \text{ expressed in m}^2/\text{kg}$$

It is derived from:

$$\left|\frac{dV}{dt}\right| = \frac{V}{2}\left|\frac{d\bar{e}}{dt}\right| = \frac{3}{4}P\sigma$$

It was shown in Section 3.4 that a two-thrust manoeuvre changes the mean longitude, even if the effect on the mean longitude drift rate is small. If the deadband is so narrow that this causes a problem, the solution is to perform a three-thrust manoeuvre according to Section 3.4. This is essentially done by splitting the first thrust in the two-thrust sequence so that part of it is performed at the same orbital position one sidereal day later. For simplicity, this split may be done in half and half, but other proportions may be found according to further optimisation criteria.

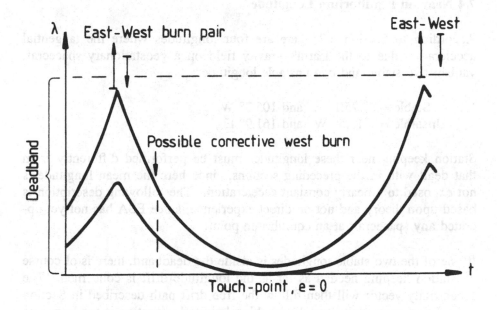

Figure 7.3.E. Schematic drawing of the longitude envelope with two-thrust station keeping. The size and time of the east-west thrust pair is aimed at reducing the eccentricity to zero at the centre of the cycle, i.e. where the longitude touches the western boundary of the deadband. If the thrust error is such that the western boundary threatens to be violated one must perform a corrective west thrust to slow down the westward drift.

The influence of thrust errors on the longitude drift rate is more severe for multi-thrust station keeping than for single-thrust methods. The reason is that the drift rate is proportional to the difference between two, possibly large, numbers, namely the east thrust $|\Delta V_1|$ and the west thrust $|\Delta V_2|$. The worst-case error in the drift rate, on the other hand, is proportional to the sum of the thrust errors. It is often impractical to subtract a longitude margin from the deadband to cover such a large error. Instead one may have to perform an extra drift correction thrust to avoid violating the deadband, Figure 7.2.E.

7.4 Near an Equilibrium Longitude

According to Section 4.2, there are four longitudes where the tangential acceleration due to the Earth's gravity field on a geostationary spacecraft vanish; two stable and two unstable longitudes:

Stable =	75.1° E	and	105.3° W
Unstable =	11.5° W	and	161.9° E

Station keeping near these longitudes must be performed differently from that dealt with in the preceding sections, since here the mean longitude is not exposed to a nearly constant acceleration. The following description is based upon theory and not on direct experience, since ESA has not yet operated any spacecraft at an equilibrium point.

If one of the two stable longitudes is inside the deadband, there is of course no station keeping needed as far as the longitude drift is concerned. The eccentricity vector will then follow the free drift path described in Section 4.5. If this eccentricity and the resulting longitude libration is too large one must reduce the eccentricity with an east-west or west-east thrust pair, or possibly three thrusts, as described at the end of the preceding section. The two thrusts will nominally be of equal size, in absolute value, so as not to introduce a longitude drift rate. The drift will, however, be disturbed by the thrust errors.

The free drift motion of the mean longitude around a stable point is that of a harmonic oscillator with a period of more than two years and no damping. The period is too long to allow the natural swing-back motion to compensate for the thrust errors. Instead, the errors have to be balanced by the next station keeping manoeuvre, the errors of which in its turn are compensated for by the next manoeuvre, and so on. These manoeuvre preparations have to be made as before by numerical integration of the predicted orbit based on the orbit determination after the manoeuvre thrusts.

Station keeping at an unstable longitude always needs active manoeuvring, but the fuel needed to correct the mean longitude drift is very small. The task of station keeping is similar to that of balancing a ball on the top of a hill, without any friction present. The acceleration of the mean spacecraft longitude in the immediate vicinity of an unstable equilibrium longitude λ_0 can be approximated by a linear function in λ with the proportionality constant k^2

$$\frac{d^2\lambda}{dt^2} = k^2 (\lambda - \lambda_0)$$

From Table 2 we can calculate $1/k$ to be of the order of 120 or 130 days for the two unstable longitude positions. The differential equation has got two partial solutions, one convergent and one divergent:

The convergent solution

$$\lambda = \lambda_0 + C_1 e^{-kt} \quad ; \quad \dot{\lambda} = -kC_1 e^{-kt} = -k(\lambda - \lambda_0)$$

The divergent solution

$$\lambda = \lambda_0 + C_2 e^{+kt} \quad ; \quad \dot{\lambda} = +kC_2 e^{+kt} = k(\lambda - \lambda_0)$$

The general solution of the differential equation is a linear combination of the two particular solutions. Whatever the starting conditions are, the evolution of λ will follow the dominating divergent solution after a time span of the order of $1/k$ or longer. Longitude station keeping can be described as attempts to manoeuvre the mean longitude to the convergent solution, whereas any small error will bring it back again to the divergent solution. The station keeping cycle will then look as follows, in a simplified form.

Assume that the mean longitude λ is allowed to deviate by $\pm \delta\lambda$ from λ_0. If λ_0 is not the centre of the deadband, the following formulae are valid with different values of $\delta\lambda$ on the two sides of λ_0. The error ε_λ in the orbit determination of λ is assumed to be small relative to $\delta\lambda$. The error in determination of the mean longitude drift rate, D, is called ε_D.

A longitude station keeping manoeuvre is due when λ is at the eastern boundary on the divergent solution

$$\lambda = \lambda_0 + \delta\lambda \quad ; \quad \dot{\lambda} = \psi D_{old} = k\,\delta\lambda$$

By performing an east thrust at time t_b

$$\Delta V = -(V/3)\,(D_{new} - D_{old}) = (A/3)\,2\,k\,\delta\lambda$$

we change the drift rate by $-2\,k\,\delta\lambda$ to move λ on to the convergent solution, Figure A:

$$\lambda = \lambda_0 + \delta\lambda \quad ; \quad \dot{\lambda} = \psi D_{new} = -k\,\delta\lambda$$

The convergent solution will bring λ back towards λ_0, but because of the error ε there will be a component of the divergent solution

$$\lambda = \lambda_0 + \varepsilon\, e^{k(t - t_b)}$$

that grows faster and brings λ back to the eastern boundary when

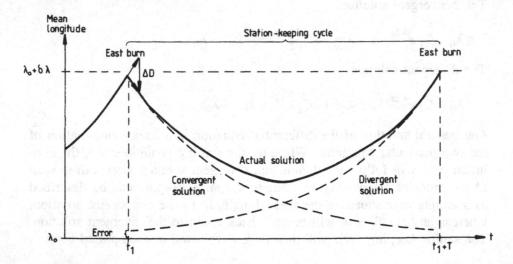

Figure 7.4.A. Schematic illustration of longitude station keeping near an unstable equilibrium longitude λ_0.

$t = t_b + (1/k) \ln(\delta\lambda/\varepsilon)$

Here ε is the root sum square of the three errors

$\varepsilon^2 = \varepsilon_\lambda^2 + (\psi^2\varepsilon_D^2 + \varepsilon_\Delta^2)/k^2$

We denote here by ε_Δ the error in the drift rate change resulting from the error in the thrust execution ΔV. Since the relative error of the thrust from a calibrated hydrazine thruster may be typically 5%, we can have $\varepsilon_\Delta = 0.005 \times 2 \times k\,\delta\lambda$. The errors could also cause λ to diverge on the western side of λ_0 but the result is analogous. In order to show a representative station keeping, we insert the typical values $\varepsilon_\lambda = k\psi\varepsilon_D = 0.1 \times \delta\lambda$ and get

$\varepsilon = 0.1 \times \sqrt{3} \times \delta\lambda$

which gives a mean station keeping cycle length of

$T = (1/k) \ln(10/\sqrt{3}) \approx 220 \text{ days}$

Because of the statistical fluctuation of the errors, the station keeping cycle will be irregular, but on the average we need to perform about two thrusts per year. The size of the thrusts is proportional to $\delta\lambda$. When $\delta\lambda = 1°$ we get ΔV's of about 0.05 m/s, so the fuel consumption is really small. The influence of these tangential thrusts on the eccentricity is negligible in comparison with the intermediate-term lunar gravity perturbation treated in Section 4.3.

The effect of librations caused by the eccentricity in the longitude around its mean value does not essentially change the above equations for the mean longitude evolution. The analytical expressions are only given for illustrative purposes, however, and the actual calculations during spacecraft operations have to be made by numerical integration of the exact orbit in each case.

The eccentricity vector will follow closely the free drift path shown in Section 4.5 for this active station keeping at an unstable equilibrium longitude in the same way as at a stable longitude without station keeping. If there are constraints on the eccentricity or the libration one must reduce the eccentricity by east-west or west-east thrust pairs. The net longitude drift change from each pair has then to be used for the active station keeping of the mean longitude, according to the scenario above. The error ε_Δ is now much greater than before since the drift rate is obtained from the difference, in absolute value, between two relatively large opposite thrusts in a thrust pair.

The active station keeping of the eccentricity vector near the stable or unstable longitude equilibria is done in an analogous way to the two-thrust sun-pointing-perigee method described in Section 7.3. The two-thrust pairs are used to move the eccentricity vector in a circle during one year, as shown in Figure 7.3.A. The radius of this circle is taken to be approximately equal to the maximum allowed eccentricity e_{max}. The fuel consumption becomes approximately

$$0.75\,P\sigma - \pi(V/Y)\,e_{max} = 108 \times \sigma - 9960 \times e_{max}\ \text{m/s/year}$$

where σ is expressed in m^2/kg. This fuel consumption is only valid when e_{max} is greater than about 7×10^{-5}, i.e. greater than the diameter of the loops of the intermediate-term lunar perturbations of Section 4.3.

As before, the target for the eccentricity station keeping preparation is to obtain a vector $\bar{e}$ of size e_{max} pointing towards the Sun at approximately the mid-point of each cycle. If the longitude is near the eastern boundary at the time when the station keeping is due, the east thrust will be performed first and the west thrust half a sidereal day later. Similarly, near the western boundary one makes the west thrust first and the east thrust afterwards. This is to ensure that the strong drift change of the first thrust moves the longitude inwards and not out of the deadband, before the second thrust has time to neutralise the drift effect.

Contrary to popular belief, there is no natural accumulation of old passive spacecraft at the stable equilibrium longitudes. Since there is no damping in the longitude drift motion one can only bring a spacecraft to an equilibrium longitude by a shift operation, as described in the next section. Theoretically one could get a spacecraft to move very slowly towards an unstable longitude and asymptotically stop by itself according to the convergent solution. Because of the influence of errors, however, one would still have to make small adjustments in accordance with the station keeping method already described. Another reason why a passive spacecraft would not remain at a stable longitude, even if it were initially manoeuvred there, is the increase in inclination described in Section 4.4. For a higher inclined orbit the along-track perturbations would no longer be the same as close to the equator plane.

7.5 Longitude Shift and Reacquisition

At the start of a geostationary mission the spacecraft is brought into the desired orbit by a series of station acquisition manoeuvres. The calculation of these manoeuvres is often more complicated than of the station keeping manoeuvres and will not be dealt with here. However, during the course of operations of a geostationary mission the spacecraft may be required to perform a simplified set of acquisition manoeuvres for the purpose of longitude shift or station reacquisition.

In general, the longitude shift task is formulated as a request for moving the spacecraft from its present longitude λ_1 to the target longitude λ_2 during a time interval T, starting on a given day. This may be part of the nominal mission or due to changes in the mission definition, as with ESA's OTS spacecraft, Figure A. There may be restrictions on the size of the eccentricity during the drift. Usually the longitude difference $|\lambda_1 - \lambda_2|$ is much greater than the station keeping deadbands at either longitude, and the mean drift rate and librations are much larger during the shift than in the station keeping cycle.

If the initial orbit has a low eccentricity one can start the shift in principle at any time by one or two tangential thrusts, as in Figures 3.3.C and 3.3.D or 3.4.A. The target mean longitude drift rate becomes:

$$\psi D = (\lambda_2 - \lambda_1)/T$$

According to Section 3.3 the total ΔV for the start thrust or thrusts shall be:

$$\Delta V = - A \, (\lambda_2 - \lambda_1)/3T$$

The ΔV to stop the drift will be approximately the same as for the start but with the opposite sign. The total ΔV for the shift is then proportional to the mean longitude drift rate by $2 \times 2.84 = 5.68$ m/s for each degree per day. By making the drift very slow one can save fuel for the shift, except that a low drift during a long time interval will be relatively more influenced by the tangential acceleration of the Earth's gravity field.

When the accuracy of the drift is important one needs to calculate the start thrusts accurately with the aid of a computer program that numerically integrates the shift orbit, starting from the actual departure orbit from the latest determination. With lower accuracy requirements one can use the simple formula above for the shift start. In either case, one shall perform further orbit determinations during the shift in order to be able to prepare an accurate drift stop inside the target deadband $\lambda_2 \pm \delta\lambda_2$.

There are essentially two different ways to perform a longitude drift:

*   With an eccentric drift orbit.

*   With a circular drift orbit.

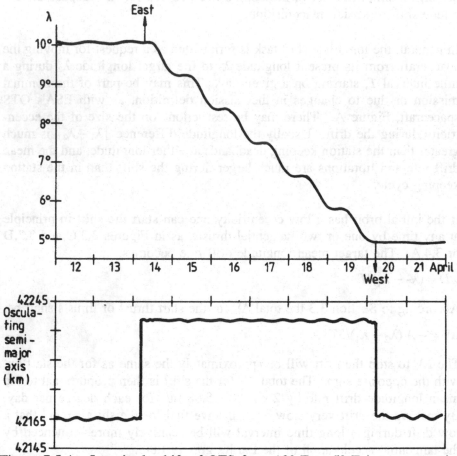

Figure 7.5.A. Longitude shift of OTS from 10° E to 5° E in an eccentric drift orbit. The drift was started by one 2.46 m/s east thrust on 1982 April 14 at 05:00 UTC and stopped by one 2.40 m/s west thrust, 6 days later at 04:58.

Eccentric drift orbit.
An eccentric drift orbit is started by one single thrust that causes the longitude to librate as shown in Figures A and B and in Figure 3.3.D. This need not cause any problems if the payload is not operated during the shift and if there is no proximity to other spacecraft. The drift should be stopped by one thrust executed at approximately the same sidereal angle as the start thrust, after an integral number of sidereal days T in order to bring the eccentricity back to a low value.

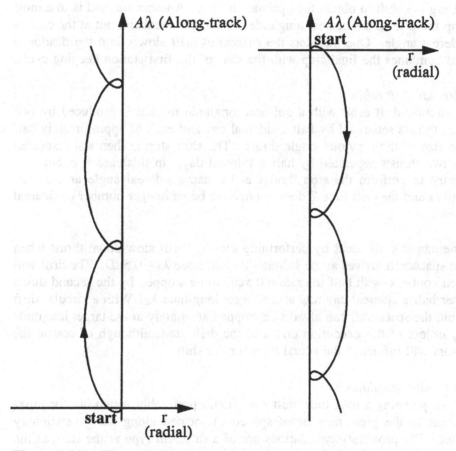

Figure 7.5.B. The radial and along-track motion of an eccentric drift orbit in the Earth-rotating coordinate system. To the left is shown an east drift, started by a single west thrust and to the right a west drift, started by an east thrust.

It is desirable to plan the drift such as to ensure that the spacecraft will be inside the target longitude deadband at the correct sidereal angle to enable the station keeping to start immediately. This plan can become invalidated by execution errors at the drift start, but one can always find another stop time at the desired angle inside the deadband if its size $2\delta\lambda_2$ is wider than the longitude drift $2\pi D$ during one orbital period $2\pi/\psi$.

If this is not the case, i.e. if $2\delta\lambda_2 < 2\pi D$ due to a high drift and a narrow deadband, one can perform a small drift correction manoeuvre half-way during the shift to obtain the optimal phasing. Another method is to almost stop the drift at the nearest longitude outside the deadband but at the correct sidereal angle. One then lets the spacecraft drift slowly into the deadband and combines the final stop with the start of the first station keeping cycle.

Circular drift orbit.
A circular drift orbit with a uniform longitude motion is produced by two start thrusts separated by half a sidereal day and each of approximately half the size of the previous single thrust. The shift stop is then also executed by two thrusts separated by half a sidereal day. In this case it is not necessary to perform the stop thrusts at the same sidereal angle as the start thrusts and the drift time T does not need to be an integer number of sidereal days.

One can stop the drift by performing the first drift slow-down thrust when the spacecraft arrives at the subsatellite longitude $\lambda_2 - 0.5\pi D$. The drift will then continue with half the rate $= 0.5\psi D$ to be stopped by the second thrust after half a sidereal day π/ψ at the target longitude λ_2. With a circular drift orbit, the spacecraft can always be stopped accurately at the target longitude regardless of the execution errors at the drift start, although of course the errors will influence the actual time for the shift.

Proximity calculations.
When planning a longitude shift one should preferably watch out for proximities to the great number of spacecraft located along the geostationary orbit. The proximity calculations are of a different type at the start, at the stop and during the shift.

For the start manoeuvre of a longitude shift to remove a member from a cluster of co-located spacecraft (Section 5.6) one can check the proximity risk with the formula given in Section 5.7. The motion of the manoeuvred

spacecraft relative to the remaining ones will be the curve that is plotted in
Figure B. Analogous calculations, although in the reverse, are made for
stopping a drifting spacecraft close to another spacecraft or inside an exist-
ing co-location cluster.

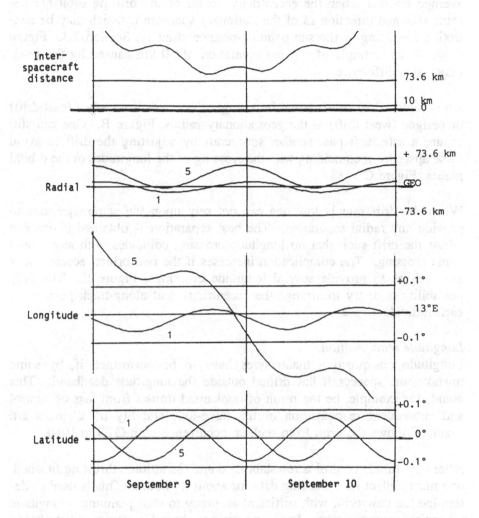

Figure 7.5.C. The west drift of ECS-5 in an eccentric orbit past ECS-1 at
longitude 13° E on 1988 September 9 and 10. The longitude crossing point
coincided with the drift orbit apogee with a wide separation in both the radial and the latitude directions. The longitude and latitude deadbands of
ECS-1 = 0.1° = 73.6 km are marked in the plot.

During the shift it is easy to drift safely past all geostationary spacecraft if the drift rate is so high that the semimajor axis is considerably higher or lower than the geostationary radius. The shift has to be started by two thrusts to provide a low eccentricity drift orbit. The radial distance is at the average optimal when the eccentricity vector of the drifting orbit has the same size and direction as of the stationary spacecraft, which may be controlled according to the sun-pointing-perigee strategy, Section 7.3. Figure D shows an example of a radial separation of 10 km caused by the semimajor axis difference.

An eccentric drift orbit started by a single thrust has the apogee (east drift) or perigee (west drift) at the geostationary radius, Figure B. One can still ensure a safe drift past another spacecraft by adjusting the drift to avoid these apses from coinciding with the crossing of the longitudes or the orbital planes, Figure C.

When the drift rate is low one can not rely upon the semimajor axis to provide any radial separation. The best separation is obtained if one can adjust the drift such that no longitude crossing coincides with any orbital plane crossing. The complication increases if the two orbital eccentricities are such as to provide several longitude crossings, Figure E. The only possibility is to try to arrange the eccentricity and along-track position in each individual case.

Longitude reacquisition.
Longitude reacquisition manoeuvres have to be performed if, by some mistake, the spacecraft has drifted outside the longitude deadband. This could, for example, be the result of unplanned thrusts from loss of control and subsequent reacquisition of the three-axis stability of a spacecraft. Figure F shows the orbit from such an occurrence with OTS in 1980.

After the attitude control is re-established and the attitude thrusting finished, one must collect new tracking data for about 24 hours. This is used to determine the new orbit with sufficient accuracy to start planning a longitude reacquisition manoeuvre. In the worst case the orbit after the disturbance has a relatively high longitude drift rate and eccentricity, as in Figure F.

There is naturally a strong requirement to move the longitude back into the deadband as fast as possible by a drift reversing tangential thrust. In order to decrease the eccentricity with the thrust one must, however, wait till the

next apogee or perigee, depending upon whether the drift is towards east
or west, respectively. If the drift reversal is performed immediately, without
regard to the eccentricity, one may later have to perform large eccentricity
reduction manoeuvres, which is difficult to do without disturbing the now
reduced longitude drift rate.

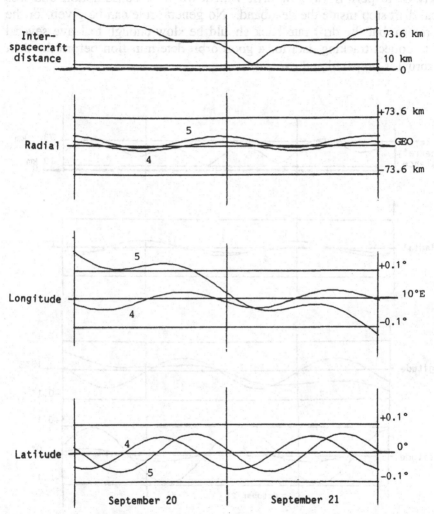

Figure 7.5.D. The slowed down west drift of ECS-5 past ECS-4 at longitude 10° E on 1988 September 20 and 21. The longitude crossing point
coincided here with the orbital plane crossing, but the radial separation
provided just the safe distance of 10 km.

In the worst case one may then have to wait for up to two days after the attitude has been re-established to perform the drift reversal manoeuvre; one day to collect tracking data and another day to wait for the suitable sidereal angle for the reversal thrust. The size of the return drift rate must then be established by the operator as a compromise between the wish to get back quickly or to perform a slow drift with lower fuel consumption and less critical drift stop inside the deadband. No general rule can be given for the drift return, but the drift rate back should be slow enough to leave several days to collect tracking data for a good orbit determination before the drift stop thrust is prepared and executed.

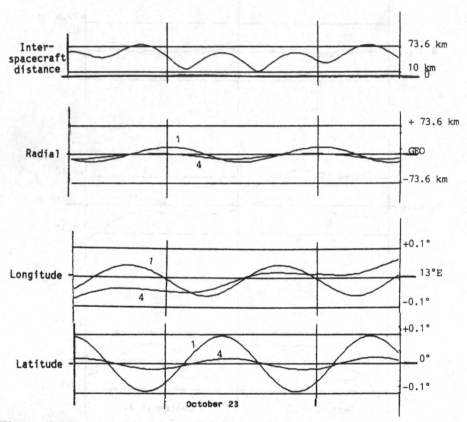

Figure 7.5.E. A very slow east drift of ECS-4 past ECS-1 at longitude 13° E with three longitude crossing points on 1988 October 23. The safe distance of 10 km was ensured in at least one of the three dimensions by differences in the inclination and eccentricity.

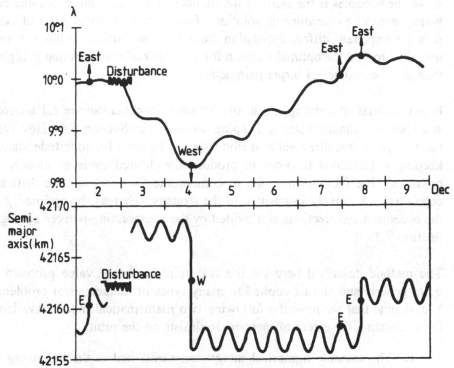

Figure 7.5.F. In the afternoon of 1980 December 2 the OTS spacecraft lost the three-axis control due to an out-of-limits attitude depointing test. This and the attitude restabilisation disturbed the orbit during 12 hours, so the longitude drifted outside the $10° \pm 0.1°$ deadband. The drift was reversed by a -0.382 m/s thrust and finally stopped by a +0.015 m/s and a +0.161 m/s thrust.

7.6 Manoeuvre Calculation

In this section we will give a brief account of the computational method normally used in computer programs for calculating the manoeuvres to be performed in order to produce a certain prescribed effect on the orbit. The mathematical question addressed here is the *two-point boundary value* problem, meaning that the initial orbit is given and a set of criteria for the target orbit is specified at a target epoch of the future. The number of solve-for variables is the same as the number of specified target parameters, which leads to a deterministic solution. From a mathematical point of view it is a completely different question from an *optimisation* problem, where one aims to find the optimal solution for a number of variables that is higher than the the number of target parameters.

Fuel optimisation tasks appear in station acquisition manoeuvre calculations and also in inclination station keeping, as shown in Section 6.4. However, the two-point boundary value method should be used for longitude station keeping calculations in order to produce the desired cycle as closely as possible. The fuel consumption to compensate for the longitude drift acceleration is relatively insensitive to the strategy, whereas the optimality of the eccentricity corrections is provided by the sun-pointing-perigee strategy, Section 7.3.

The method described here for the two-point boundary value problem is quite general and can be applied to many types of mathematical problems. We assume that we have the following two mathematical models available for calculating the effect of manoeuvre thrusts on the orbit:

- 1. The accurate numerical integration mentioned at the beginning of Section 3.1 to propagate an orbit that is influenced by the combined effect of a sequence of thrusts (long or short) and of the natural perturbations.

- 2. The simplified approximate relations of Sections 3.3 and 3.4 that describe the linearised effect of one or more instantaneous ΔVs on an unperturbed orbit.

The problem is now caused by the fact that only model 2 above, but not 1, can be inverted in a simple way to calculate which ΔVs or accelerations shall be applied to produce the desired effect on the orbit. On the other hand, model 2 may not be accurate enough for the operational requirements, so by preference model 1 should be used. We will show here how this can be solved by combining the two models into an iterative process.

To start with we describe here a very general iterative scheme that is applicable not only to orbits. For this purpose we introduce here the following mathematical notation:

- $x =$ a vector consisting of the parameters to be solved. This can be a combination of one or more tangential ΔVs and their respective execution times. It could also be a series of switch on and off times for long thrust durations.

- $y =$ a vector consisting of the orbit criteria that appear in the target. This can be any combination of orbital elements or other properties at a specified epoch or time interval in the future. It must have the same dimension as x.

- $Y =$ the specified target value of y.

- $F(x) = y =$ the effect on the orbit criteria at the specified target epoch or time interval of the manoeuvres x according to the accurate model 1 above. Its inverse cannot be calculated.

- $f(x) =$ an approximation of $F(x)$ according to model 2. The inverse $x = f^{-1}(y)$ can be calculated directly.

- $X =$ the desired solution of x satisfying $F(X) = Y$.

- $x = 0 =$ the vector x for the case without manoeuvres.

The iteration for solving $F(X) = Y$ consists of calculating a series of approximations x_n to X as follows:

$$x_{n+1} = f^{-1}[f(x_n) + Y - F(x_n)] \quad \text{starting with } x_0 = 0$$

The iteration can be expressed in the following words. One propagates the orbit by the accurate integration $F(x_n)$ to the target epoch with the manoeuvres from the latest iteration step and subtracts it from the target Y. The difference between the two is the current iteration error. This error is added to the result of the approximate propagation $f(x_n)$ and the whole is inserted into the inverse f^{-1} of the approximate model. The result is the set of manoeuvres to be used for the next iteration step. The first step is performed in the same way, except that the orbit is propagated without manoeuvres.

The iteration is considered to have converged if either $x_{n+1} - x_n$ or $Y - F(x_n)$ becomes sufficiently small. The two criteria are equivalent provided that the function $f(x)$ is continuous and has a unique and continuous inverse. Sometimes, when $f(x)$ has a periodic part, one may have to limit the time interval allocated for the manoeuvres in order to obtain a unique inverse. As a rule of thumb one can allow the remaining error to be as small as the physical error in the thrust performance or in the orbit determination.

The accuracy of the found solution X is not directly influenced by how accurate the approximation $f(x)$ is. Only the number of iterations necessary for convergence is affected. However, if it is very inaccurate it may happen that the iteration does not converge. It is not possible to prove by mathematical means that this iteration scheme converges in the general case, but in many practical applications it has been found to converge by trial.

The iteration can be simplified for the special case when the approximate model $f(x)$ is linear. This is typically the case when the vector x of parameters to be solved only contains the ΔVs but not the thrust times, as we will see below. The linear iteration becomes:

$$x_{n+1} = x_n + f^{-1}[Y - F(x_n)] \quad \text{starting with} \quad x_0 = 0$$

The following practical examples show the application of the above or analogous methods to solve problems with 1, 2 or 3 variables for geostationary longitude manoeuvres.

Example 1. An impulsive thrust to a given longitude.
Task: Find an impulsive ΔV at a given time t_b such that the longitude $\lambda(t_T)$ at a specified future epoch t_T has the target value λ_T.

The accurate model $F(\Delta V_n)$ is obtained by adding ΔV_n to the velocity part of the spacecraft state vector at time t_b, integrate it to t_T and there calculate the subsatellite longitude $\lambda_n(t_T)$. The approximate model is linear, according to Section 3.3:

$$f(\Delta V_n) = \Delta V_n[- (3/A)(t_T - t_b) + (4/V) \sin \psi(t_T - t_b)]$$

In order to ensure that the function $f( )$ does not become too small for being inverted, one has to prescribe the time interval $t_T - t_b$ to be greater than, e.g. one day. If the interval is longer than a few days one can also use the even simpler approximation:

$$f(\Delta V_n) = - \Delta V_n(3/A)(t_T - t_b)$$

The iteration becomes in the latter case:

$$\Delta V_{n+1} = \Delta V_n + \frac{A}{3} \frac{\lambda_n(t_T) - \lambda_T}{t_T - t_b} \qquad \text{starting with } \Delta V_0 = 0$$

Example 2. A long weak thrust to a given longitude.
Task: Find the duration Δt of a given weak thrust with acceleration $= H$, starting at a given time t_1 such that the longitude $\lambda(t_T)$ at a specified future epoch t_T has the target value λ_T.

This is similar to Example 1 but with a long thrust. The accurate model $F(\Delta t_n)$ is obtained by integrating the spacecraft state vector from t_1 to t_T, of which the first part Δt includes the thrust acceleration. There are several possible iteration schemes, of which three are given below. The first one is suitable when Δt is relatively long, whereas the last one can be used when $t_T - t_1$ is longer than several days. In all cases one starts with $\Delta t_0 = 0$.

$$\Delta t_{n+1} = \Delta t_n + \frac{V}{H} \frac{\lambda_n(t_T) - \lambda_T}{3\psi(t_T - t_1 - 0.5\Delta t_n) - 4 \sin \psi(t_T - t_1 - 0.5\Delta t_n)}$$

$$\Delta t_{n+1} = \Delta t_n + \frac{V}{H} \frac{\lambda_n(t_T) - \lambda_T}{3\psi(t_T - t_1) - 4 \sin \psi(t_T - t_1)}$$

$$\Delta t_{n+1} = \Delta t_n + \frac{A}{3H} \frac{\lambda_n(t_T) - \lambda_T}{t_T - t_1}$$

Example 3. An impulsive thrust to obtain a longitude that touches a given longitude from east.
Task: Find an impulsive ΔV at a given time t_b such that the resulting longitude $\lambda(t)$ during a specified future time interval $t_L \leq t \leq t_M$ stays east of but touches the target longitude λ_T, as follows:

$$\min_{t_L \leq t \leq t_M} \lambda(t) = \lambda_T$$

This method is used for calculating the longitude station keeping manoeuvre shown in Figure 7.2.B, where λ_T is the western deadband boundary. The same method can be used for touching the eastern deadband boundary from west by swapping the minimum for the maximum. The accurate model, for the minimum case, becomes:

$$F(\Delta V_n) = \min_{t_L \leq t \leq t_M} \lambda_n(t)$$

where the longitude $\lambda_n(t)$ is the result of integrating the determined orbit with the actual ΔV_n from the latest iteration. For the approximate model one can take the linear function, as in Examples 1 and 2. However, the time t_m below must be the same time as where the minimum is attained in the above equation, which varies with the iteration step.

$$f(\Delta V_n) = \Delta V_n[-(3/A)(t_m - t_b) + (4/V)\sin \psi(t_m - t_b)]$$

One can avoid calculating this by using the modified iteration scheme below, which is also slightly more efficient. In order to avoid a singularity in the denominator one must restrict the lower end of the time interval to $t_L \geq t_b + 1$ sidereal day.

$$\Delta V_{n+1} = \Delta V_n + \min_{t_L \leq t \leq t_M} \frac{V(\lambda_n(t) - \lambda_T)}{3\psi(t - t_b) - 4\sin\psi(t - t_b)} \quad \text{starting with} \quad \Delta V_0 = 0$$

Example 4. Calculate the thrust as in Example 3 but find a thrust time to change the eccentricity vector as close as possible to a given target.
Task: There are two variables: ΔV and t_b. The longitude criterion is the same as in Example 3, but in addition the mean eccentricity vector $\bar{e}(t_T)$ at a future epoch t_T shall be manoeuvred as close as possible to a target vector $\bar{e}_T$.

The two parameters are calculated in the same iteration with a similar solution for ΔV as in Example 3. Since ΔV is prescribed by the longitude criterion, there is only one variable, the thrust time, left to satisfy the eccentricity vector target. The mean eccentricity vector found in the solution will be the point in the eccentricity vector plane closest to the target that can be reached with the given ΔV.

The accurate model is the numerical integration that starts with the latest value ΔV_n, t_n of the iterated variables thrust size and time, to produce an orbit with $\lambda_n(t)$ and $\bar{e}_n(t_T)$. The approximate model is non-linear. The effect on the eccentricity is periodic, so the program has to restrict the solve for times t_n to one specific orbital revolution $t_0 \leq t_n < t_0 + 2\pi/\psi$, where t_0 is an operator input. We define the corresponding spacecraft sidereal angle to be $s_n = s_0 + \psi(t_n - t_0)$. The combined iteration becomes:

$$\Delta V_{n+1} = \Delta V_n + \min_{t_L \leq t \leq t_M} \frac{V(\lambda_n(t) - \lambda_T)}{3\psi(t - t_n) - 4\sin\psi(t - t_n)} \quad \text{starting with} \quad \Delta V_0 = 0; \ t_0$$

$$C \begin{pmatrix} \cos s_{n+1} \\ \sin s_{n+1} \end{pmatrix} = \frac{2\Delta V_n}{V} \begin{pmatrix} \cos s_n \\ \sin s_n \end{pmatrix} + \bar{e}_T - \bar{e}_n(t_T) \qquad \text{starting with } s_0$$

In the formula above, C is an arbitrary scalar that must have the same sign as ΔV_{n+1} in order to produce the correct solution for s_{n+1}. We solve the latter with an arctangent function of the right hand side and adjust it to the interval $s_0 \leq s_{n+1} < s_0 + 2\pi$. The thrust time then becomes

$$t_{n+1} = t_0 + (s_{n+1} - s_0)/\psi$$

Example 5. Find two thrusts and a time to reach a given eccentricity vector and a longitude as in Example 4.
Task: This problem contains the three variables ΔV_1, ΔV_2, t_b. The longitude criterion is the same as in Examples 3 and 4, but the mean eccentricity vector $\bar{e}(t_T)$ at a future epoch t_T shall be equal to a target vector $\bar{e}_T$.

The two thrusts shall be executed half a sidereal day apart, so only one time appears among the free variables. We define the execution time t_b of ΔV_1 to be restricted to lie inside the half sidereal day $t_0 \leq t_b < t_0 + \pi/\psi$. The time for ΔV_2 then becomes $t_b + \pi/\psi$. The accurate model is analogous to Example 4.

For the iteration we define two auxiliary parameters and the spacecraft sidereal angle:

$$U = \Delta V_1 + \Delta V_2 \quad ; \quad W = \Delta V_1 - \Delta V_2$$

$$s_n = s_0 + \psi(t_n - t_0)$$

The new combined iteration becomes:

$$U_{n+1} = U_n + \frac{V}{3} \min_{t_L \leq t \leq t_M} \frac{\lambda_n(t) - \lambda_T}{\psi(t - t_n) - 0.5\pi} \qquad \text{starting with } U_0 = 0$$

$$W_{n+1} \begin{pmatrix} \cos s_{n+1} \\ \sin s_{n+1} \end{pmatrix} = W_n \begin{pmatrix} \cos s_n \\ \sin s_n \end{pmatrix} + \frac{V}{2} [\bar{e}_T - \bar{e}_n(t_T)] \quad \text{starting with } W_0 = 0; \; s_0$$

Before solving W_{n+1} above we must adjust s_{n+1} to lie in the interval $s_0 \leq s_{n+1} < s_0 + \pi$. We solve it with an arctangent function and obtain the thrust time by $t_{n+1} = t_0 + (s_{n+1} - s_0)/\psi$.

Example 6. A thrust to produce a longitude that is symmetric around a given longitude.

Task: Find an impulsive ΔV at a given time t_b such that the resulting longitude $\lambda(t)$ during a specified future time interval $t_L \leq t \leq t_M$ has its maximum and minimum symmetrically around a target longitude λ_T. This method is used for calculating the longitude station keeping manoeuvre shown in Figure 7.2.C, where λ_T is the centre of the deadband. By introducing the notation W for the symmetric longitude deviation on either side of λ_T one can write the target condition as:

$$\lambda_T - \min_{t_L \leq t \leq t_M} \lambda(t) = \max_{t_L \leq t \leq t_M} \lambda(t) - \lambda_T = W$$

The longitude criterion is analogous to Example 3 except for the additional parameter W. It can be combined with a solve-for thrust time to approach a target eccentricity as in Example 4 and also be extended to two thrusts as in Example 5. As before we denote the accurate model $\lambda_n(t)$ to be the longitude that results from integrating the orbit with the actual ΔV_n of the current iteration step. The approximate model is the same linear function as before. To facilitate the writing of the equations we introduce temporarily the function:

$$g(t) = (3/A)(t - t_b) - (4/V) \sin \psi(t - t_b)$$

In each step we have to solve also for W in the iteration. We introduce the two new variables U_A, U_B through the following equations:

$$U_A = \min_{t_L \leq t \leq t_M} \frac{\lambda_n(t) - \lambda_T + W_n}{g(t)} \qquad \text{where the minimum is attained at } t = t_A$$

$$U_B = \max_{t_L \leq t \leq t_M} \frac{\lambda_n(t) - \lambda_T - W_n}{g(t)} \qquad \text{where the maximum is attained at } t = t_B$$

In the subsequent equations we will use the above expressions with the values of t inserted that gives the maximum and minimum, respectively, in the form:

$$g(t_A)U_A = \lambda_n(t_A) - \lambda_T + W_n \quad ; \quad g(t_B)U_B = \lambda_n(t_B) - \lambda_T - W_n$$

The two values U_A and U_B are equivalent to the expression in Example 3 for calculating ΔV_{n+1} in each iteration step. In this case the two values correspond to the longitude condition at the minimum and maximum, respectively, and they shall become equal after longitude symmetry has been obtained. In order to approach this symmetry during the iteration, we also solve for W_{n+1} during the same step by requiring that the two right hand sides below are equal:

$$\Delta V_{n+1} - \Delta V_n = \frac{\lambda_n(t_A) - \lambda_T + W_{n+1}}{g(t_A)} = \frac{\lambda_n(t_B) - \lambda_T - W_{n+1}}{g(t_B)}$$

In the equation above we insert U_A and U_B from the preceding equations and solve the two unknowns ΔV_{n+1} and W_{n+1}. This leads to the following formula for each iteration step:

$$\Delta V_{n+1} = \Delta V_n + \frac{g(t_A)U_A + g(t_B)U_B}{g(t_A) + g(t_B)} \qquad \text{starting with } \Delta V_0 = 0$$

$$W_{n+1} = W_n + (U_B - U_A)\frac{g(t_A)\, g(t_B)}{g(t_A) + g(t_B)} \qquad \text{starting with } W_0 = 0$$

8. ORBIT DETERMINATION

8.1 Ground Station Visibility

Instead of the mean equatorial geocentric co-ordinate system of date, MEGSD, of Section 2.1, we will here express the spacecraft position in the geocentric equatorial Earth-rotating co-ordinate system, identified by the index E, like in Section 4.2. The z-axis of this system coincides with the z-axis of MEGSD. The x-axis goes through the zero-meridian and is rotated with respect to the MEGSD x-axis by the Greenwich sidereal angle G, as shown in Figure 2.1.C.

$$x_E = x \cos G + y \sin G \quad ; \quad y_E = y \cos G - x \sin G$$

$$\bar{r}_E = \begin{pmatrix} x_E \\ y_E \\ z \end{pmatrix} = \begin{pmatrix} r \cos \theta \cos \lambda \\ r \cos \theta \sin \lambda \\ r \sin \theta \end{pmatrix}$$

The spacecraft orbit is calculated from its measured position relative to the ground stations used for tracking. The position of those and other stations used for the mission must then be known in the Earth-rotating co-ordinate system. Ground station co-ordinates can be measured by a satellite-based navigation system, like USA's military "Global Positioning System" GPS, which also has a less accurate but generally available civilian signal. This lower accuracy can be compensated for by collecting data from GPS during several days to determine the position of a ground station with an accuracy better than a metre in the corresponding co-ordinate system.

Accuracies better than a metre cannot be achieved with a model that considers the Earth to be a solid body. The same effect that causes the tidal motion of the oceans (Section 4.3), also moves the surface of the Earth with the amplitude of up to 23 cm. In addition, the continental drift is a few centimetres per year and local crustal motion may be even higher. Here we will not deal with geodetic missions, so we will be content to consider the Earth as solid with station position accuracies of the order of metres. The co-ordinate system definition uses an idealised shape of the Earth,

which is an ellipsoid of rotation with *equatorial radius R* = 6378.144 km
and *oblateness f* = 1/298.257, Figure A.

Ground station positions are given in this system as geodetic *latitude, (φ),*
longitude (l) and *height (h)* above the geoid. The local horizontal plane is
parallel to the tangent of the ellipsoid at the station and orthogonal to the
zenith direction. Because of the Earth's oblateness the zenith direction, and
its opposite direction, *nadir*, do not pass through the centre of the Earth.
The deviation angle between the zenith and the Earth-centre-to-station direction varies with the latitude. The maximum value of 0.2° is obtained
when φ = 45° and it vanishes at the pole and the equator.

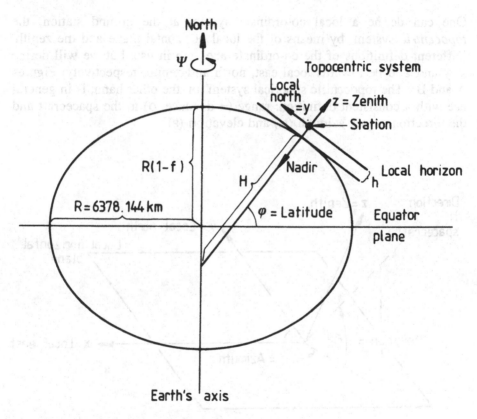

Figure 8.1.A. The reference geoid is an ellipsoid of rotation around the
Earth's north-south axis. The plane of the paper coincides with a plane
through this axis and a ground station. The oblateness f = 1/298.257 is ex-
aggerated in this picture for clarity.

The geodetic latitude ϕ of a ground station is defined as the angle between its zenith direction and the equatorial plane, Figure A. By introducing the auxiliary parameter

$$H = R/\sqrt{1 - f(2 - f)\sin^2\phi}$$

we can express the station position vector in the Earth-rotating Cartesian co-ordinates by means of (h, l, ϕ)

$$\overline{R} = \begin{pmatrix} (H + h)\ \cos\phi\ \cos l \\ (H + h)\ \cos\phi\ \sin l \\ [(1 - f)^2 H + h]\ \sin\phi \end{pmatrix}$$

One can define a local co-ordinate system at the ground station, the *topocentric* system, by means of the local horizontal plane and the zenith. Different definitions of the co-ordinate axes are in use, but we will define x, y and z to be towards local east, north and zenith, respectively, Figures A and B. The topocentric spherical system, on the other hand, is in general use with a consistent definition: range (= distance, ρ) to the spacecraft and the direction angles azimuth (α) and elevation (ϵ).

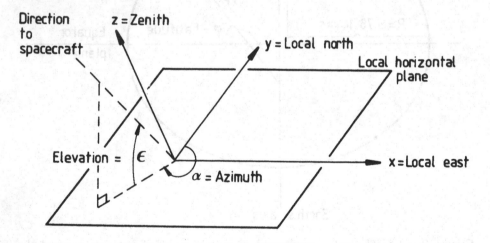

Figure 8.1.B. The topocentric co-ordinate system with ground antenna azimuth and elevation.

The elevation is the angle between the station-spacecraft direction and the horizontal plane, Figures B and C. Azimuth gives the direction of the projection of the station-spacecraft line on the horizontal plane. For historical reasons it is counted clockwise (= negative rotation) from local north, Figure B.

Large station antennas are often mounted in such a way that they can be turned around two axes. A vertical axis turns the antenna support mechanism through the azimuth angle, Figure C. On this mechanism is mounted a horizontal axis that tips the antenna to the desired elevation. Other types of antenna support mountings make the calculation of the direction angles more complicated. It is desirable that the vertical and horizontal axes intersect in one pivot point, which is the reference point to be used in the definition of the ground-station co-ordinates.

The lower limit of the elevation that is possible for station-spacecraft contact depends on the profile of the horizon around the station and on the signal attenuation of the atmosphere. When there is contact one says that the station can "see" the spacecraft, or that there is station or spacecraft "visibility".

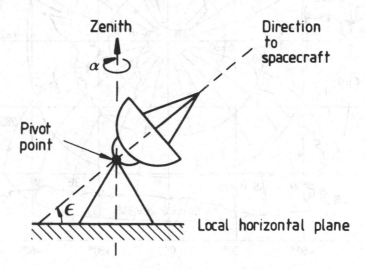

Figure 8.1.C. Schematic picture of a ground antenna that can be turned around the azimuth and elevation angles through a pivot point.

We can now obtain a relation between the spacecraft co-ordinates in the topocentric and geocentric systems by means of the station co-ordinates (h, l, ϕ). We denote by $\bar{\rho}$ the station-spacecraft vector in the topocentric Cartesian system and express it in $(\rho, \varepsilon, \alpha)$. This must be equal to $\bar{r}_E - \bar{R}$ rotated from the equatorial to the topocentric system by the matrix below:

$$\bar{\rho} = \begin{pmatrix} \rho \cos \varepsilon \sin \alpha \\ \rho \cos \varepsilon \cos \alpha \\ \rho \sin \varepsilon \end{pmatrix} = \begin{pmatrix} -\sin l & \cos l & 0 \\ -\cos l \sin \phi & -\sin l \sin \phi & \cos \phi \\ \cos l \cos \phi & \sin l \cos \phi & \sin \phi \end{pmatrix} (\bar{r}_E - \bar{R})$$

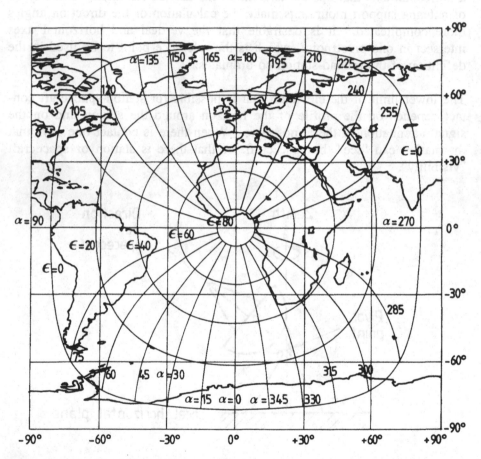

Figure 8.1.D. Imposed on the map are the contour lines of constant azimuth (α) and elevation (ε) from different locations on the Earth's surface to a geostationary spacecraft at 0° longitude.

By combining this formula with the expressions for $\bar{r}_E$ and $\bar{R}$ we can obtain the spacecraft position in the spherical topocentric co-ordinates $(\rho, \varepsilon, \alpha)$ from (r, λ, θ) and the (h, l, ϕ) or vice versa. The formulas are straightforward but tedious and will not be given here. It is more instructive to give the approximate relations for a spherical Earth $(f = 0)$ with a ground station on its surface $(h = 0)$. This approximation gives angle errors smaller than $0.2°$, which is sufficient for mission analysis purposes. The more accurate formulas can be left for computer calculations that support spacecraft operations.

To start with we assume a spacecraft that is perfectly geostationary $(\theta = 0)$ at the longitude λ. We can then calculate $(\rho, \varepsilon, \alpha)$ from

$$\rho = \sqrt{r^2 + R^2 - 2rR \cos(l - \lambda) \cos \phi} = \sqrt{r^2 - R^2 \cos^2\varepsilon} - R \sin \varepsilon$$

$$\rho \sin \varepsilon = r \cos(l - \lambda) \cos \phi - R$$

$$\rho \cos \varepsilon = r\sqrt{1 - \cos^2(l - \lambda) \cos^2\phi}$$

$$\sin \alpha = - r \sin(l - \lambda)/\rho \cos \varepsilon$$

$$\cos \alpha = - r \cos(l - \lambda) \sin \phi/\rho \cos \varepsilon$$

The following table shows the range distance ρ as a function of the elevation ε.

ε	ρ
0°	41680 km $= \sqrt{r^2 - R^2}$
10°	40587 km
20°	39555 km
30°	38612 km
40°	37781 km
50°	37079 km
60°	36520 km
70°	36115 km
80°	35869 km
90°	35787 km $= r - R$

Figure D shows part of a world map with contour lines in (l, ϕ) corresponding to constant ε or α, respectively. The spacecraft is here located at longitude $\lambda = 0°$. The contours of constant ε are actually circles on the Earth's surface, but appear distorted in this rectangular projection.

Figure E shows the Earth as it is seen from a geostationary spacecraft. The Earth's radius is seen as the angle = arcsin $R/r = 8.7°$. The lines of constant ε would be concentric circles around the subsatellite point $\lambda = 0°, \theta = 0°$. The radius of such a circle corresponds to an arc on the Earth's surface marked by ϑ in Figure 8.3.B. It is the angle, from the centre of the Earth, between the spacecraft and the station and can be obtained from

$$\vartheta = \arccos[(R/r)\cos\varepsilon] - \varepsilon = \arccos[\cos(l - \lambda)\cos\phi]$$

Figure 8.1.E. The Earth as seen from a geostationary spacecraft at 0° longitude. The Earth's diameter is seen at an angle of 17.4°.

8.2 Tracking Measurements

In this context it is of interest to distinguish between ground antennas used for spacecraft tracking and other antennas, which may be used for commanding or reception of payload data. A payload antenna can be allowed to be unmovable if its beamwidth is wider than the deadband box of the spacecraft, seen from the station. Other antennas must be steerable in autotrack or programtrack mode.

Here we must observe that, according to current terminology, *"track"* can mean two different things: measurement of the spacecraft position relative to the station or the steering of the station antenna to follow the spacecraft motion. We will use it only in the former sense, except in the combinations autotrack and programtrack.

An antenna is steered in *autotrack* mode by directly feeding the received signal into the antenna control loop to make it follow the detailed motion of the spacecraft. In *programtrack* mode the antenna is steered according to predicted spacecraft directions obtained from the orbit determination and prediction process in the control centre that are transmitted to the station in compressed form.

Spacecraft tracking data usually consists of *range* and/or *antenna angles* measured from one or more ground stations at discrete time points. Range rate or Doppler tracking is of less interest for geostationary orbits. Because of the influence of the Earth's atmosphere, one normally tries to obtain the tracking data from stations that see the spacecraft at an elevation of at least 20°. Figure A shows a sample of a refraction model of the troposphere that is also valid for visible light.

The ranging delay due to the Earth's ionosphere can be modelled by the following expression, in metres

$$\delta\rho = 40.3 \, N/f^2 \sin \varepsilon$$

where N is the integrated electron content in m^{-2}. It usually varies between 10^{16} and 10^{17} at local night and between 10^{17} and 10^{18} at local day time, with occasional peaks at 10^{19}. The carrier frequency is here denoted by f, in Hz. Typical values for communications missions are of the order of 10^9 (= 1 GHz) to 10^{10}.

Ranging measures the distance (ρ) by means of the travel time of a signal from the station to the spacecraft *(up-link)* and back *(down-link)* via the on-board transponder. It can be done in a *tone ranging* system by amplitude modulation of the carrier signal by a series of "tones" of different wavelengths. More modern systems employ a *pseudo-random code* for matching the received with the transmitted signal. The resulting range measurement then contains an unknown integral multiple of the longest tone wavelength or code length, respectively. The *ambiguity resolution* in the orbit determination program, to insert a multiple of the longest step, seldom causes any problem for a geostationary spacecraft, whose distance to a station undergoes only small variations.

During the ranging, the equipment performs many measurements for a few seconds or minutes, which are then averaged to one single ranging measurement. Corrections must be applied for delay in signal travelling time through the on-board and station electronics and for delay due to the Earth's troposphere and ionosphere. The two latter effects usually contribute to the order of several metres and sometimes even higher. The ranging accuracy is normally a few metres, but typical residuals in ranging data are of the order of 10 or 20 m because of errors in modelling the range corrections and also in the orbit propagation.

The antenna angles *elevation* (ε) and *azimuth* (α) can be obtained from a down-link autotrack ground antenna with narrow beamwidth. The angle read-off does not interfere with the payload data transmission, so large volumes of angle tracking are often available. From the measured elevation angle one must subtract the refraction of the Earth's atmosphere (Figure A) to convert it to the geometric elevation for the orbit determination. Since the refraction models in use are spherically symmetric, there is no correction to be applied to the azimuth measurement.

For carrier frequencies of the order of GHz and higher, the accuracies in the measured ε and α are limited by the mechanical stability of the antenna support structure over a time span equal to the time between calibrations. Antennas that are built for providing angle measurements have reflectors of typically 10 m to 15 m and are often specified to provide errors below 0.01°.

Since tracking measurements are carried out in the Earth-fixed coordinate system one must apply a transformation with the Earth's nutation, Section

2.1, when the orbit determination is carried out in the mean system of date (MEGSD). Depending on the accuracy requirements, one may have to consider also the polar motion and the non-uniform Earth rotation with time correction.

Tracking measurements are usually obtained on ground although one could in principle design a mission with a reverse ranging system where land-based transponders receive and return signals from the spacecraft. General navigation systems, like the Global Positioning System GPS of Section 8.1, are suitable for spacecraft in low orbits, but its signal can hardly be received in geostationary orbit with the present design.

There exist proposals to navigate by on-board sensors that measure the directions to stars, the Sun, the Earth and, possibly, the Moon. However, all the celestial bodies except the Earth appear to be at infinite distance and provide only directional information that help to define the orientation of the spacecraft. The only information of the spacecraft position that can be obtained is due to the variation in the Earth's direction, seen from the spacecraft, during the orbital motion.

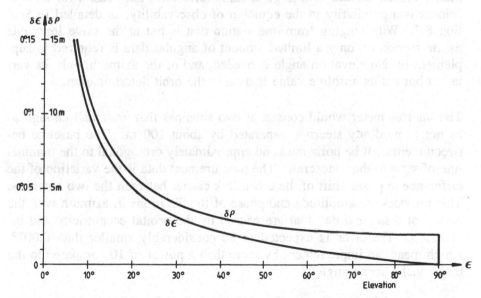

Figure 8.2.A. Example of a tropospheric refraction model. From the measured range value one must subtract $\delta\rho$ (for each pass through the atmosphere) and from the elevation $\delta\epsilon$.

Available methods to measure the Earth's direction from a spacecraft provide an even lower accuracy than measurements of the direction to the spacecraft from a ground station antenna. This is in conflict with the trend towards more stringent requirements on control accuracy with the increased exploitation of the geostationary orbit. The Earth's direction can be measured by infrared sensors, horizon scanners or land-mark trackers or obtained from the occultation of stars by the Earth or the refraction of the apparent star position by the Earth's atmosphere.

The most important application of on-board orbit determination for geostationary missions would be for spacecraft clusters with inter-satellite tracking. The types of tracking of interest are ranging and Doppler, whereas pointing angle tracking would require more complex on-board equipment. Such a design would be of importance to co-locate several spacecraft in a narrow deadband without collision risk.

A new, promising, ground tracking method that has not yet been tried in practice is to measure the azimuth angle to the spacecraft by an interferometer instead of a large antenna reflector. The reason for its usefulness is a peculiarity in the equation of observability, as detailed in Section 8.7. With ranging from one station that is not at the same longitude as the spacecraft, only a limited amount of angular data is required to supplement it: No elevation angle is needed, and of the azimuth, only its variation but not its absolute value is used in the orbit determination.

The interferometer would consist of two antennas that are small enough so as not to need any steering, separated by about 100 m. The baseline between them shall be horizontal and approximately orthogonal to the nominal line-of-sight to the spacecraft. The measurement data is the variation of the difference in phase shift of the down-link carrier between the two antennas. This provides the amplitude and phase of the librations in azimuth with the period of a sidereal day that are caused by the orbital eccentricity and inclination. The error is expected to be considerably smaller than 0.001°, which means an improvement by more than a power of 10 compared to the traditional antenna angle.

All signals from the spacecraft, including tracking measurements, are *time-tagged* when they are received at the ground station. They could have been time-tagged on-board the spacecraft by means of its own clock, but the accuracy is normally very low. High precision space-qualified clocks are too expensive and heavy to fly, except on a few selected satellites like those used for the GPS service.

The result is that each measurement point has attached its arrival time according to the station clock, which is synchronised with UTC. For very accurate measurements one would need to subtract from it the time for the down-link signal to travel from the spacecraft to the station with almost the speed of light, apart from the refraction in the atmosphere. The trajectory, in an inertial coordinate system, of a signal between a moving spacecraft and its ground station is described by the *light-time* equation.

The internationally adopted value for the speed of light $= c = 299792.458$ km/s. For a geostationary spacecraft the distances that are listed in the preceding section give rise to signal travelling times of between 0.12 and 0.14 seconds. During this time the Earth rotates by between $0.0005°$ and $0.0006°$, and the spacecraft moves about 0.4 km in the inertial coordinate system MEGSD.

Figure B shows that the ground antenna angles obtained from the down-link signal deviate by about the same angle, when measured in MEGSD, compared to the purely geometric angles in the Earth-fixed system. However, with the low accuracy of the antenna angle tracking that is available at present, it is hardly necessary to correct for the signal flight time effect in the tracking model for the orbit determination.

With ranging measurements some more care is required, since we aim to obtain an accuracy of the order of metres. We will show here that the effect on the range almost disappears because the signal travels both from the station to the spacecraft and then back. There is a certain time delay of the order of microseconds between the reception and re-transmission of the signal by the spacecraft. This on-board transponder delay is measured before the flight and is usually assumed to be constant during the mission. It is subtracted from the range measurement together with a similar delay in the ground electronics.

Apart from this correction, we consider in the following the reception and re-transmission of the signal by the spacecraft to be simultaneous. We introduce the following notations for the three relevant times.

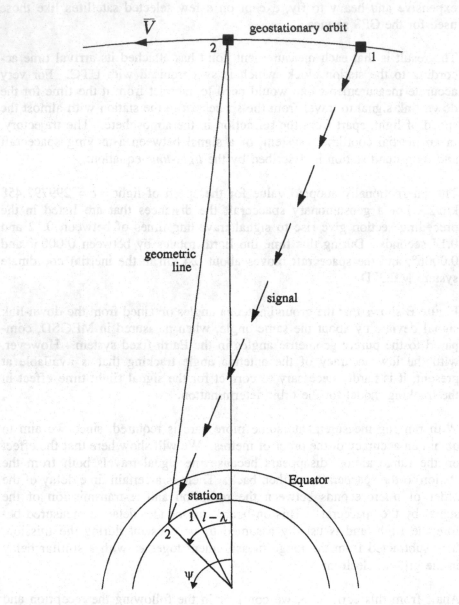

Figure 8.2.B. Schematic drawing, not to scale, in the inertial MEGSD system of the geometry for angle tracking seen from north. The down-link signal leaves the spacecraft at the position marked (1) and is received at the ground station at (2).

t_1 = time when the ground station transmits the up-link signal
t_2 = time when the spacecraft receives and re-transmits the signal
t_3 = time when the ground station receives the down-link signal

What is actually measured on ground are the two times t_1 and t_3, whereas t_2 is not apriori known. We then define the measured range as half the distance the signal travels between the transmission and reception at the station:

$$\rho_m = c\,(t_3 - t_1)/2 = (\rho_u + \rho_d)/2$$

We here denote the actual, but unknown, up-link and down-link travel distances by:

Up-link: $\rho_u = c\,(t_2 - t_1)$; Down-link: $\rho_d = c\,(t_3 - t_2)$

During the up and down transmissions, respectively, the Earth, with the station on its surface, rotates by the following two angles that both are of the order of 10^{-5} radians:

Up-link: $\delta s_u = \psi(t_2 - t_1) = \rho_u \psi/c$; Down-link: $\delta s_d = \psi(t_3 - t_2) = \rho_d \psi/c$

Figure C shows the position of a geostationary spacecraft at the time t_2 and the station at each of the three times. The purely geometric range ρ in the Earth-fixed system is the theoretical distance to the station at its position at time t_2, which was given in the previous section:

$$\rho = \sqrt{r^2 + R^2 - 2rR \cos\phi \cos(l - \lambda)}$$

The geometric calculation of the actual up-link and down-link ranges, however, must take into consideration the two small angles of rotation of the Earth between the begin and end of the transmissions, Figure C:

$$\rho_u = \sqrt{r^2 + R^2 - 2rR \cos\phi \cos(l - \lambda - \delta s_u)}$$

$$\rho_d = \sqrt{r^2 + R^2 - 2rR \cos\phi \cos(l - \lambda + \delta s_d)}$$

Here one could insert the above dependence of δs_u and δs_d on ρ_u and ρ_d, but the equations would be too difficult to solve. Instead we use the fact that the two angles are very small so we can, with a high accuracy, approximate the above expressions by the expansion to the first order in the angles. The error is then of the order of $\delta s^2 = 10^{-10}$ radians.

$$\rho_u \approx \rho - (rR/\rho) \cos \phi \sin(l - \lambda) \, \delta s_u$$

$$\rho_d \approx \rho + (rR/\rho) \cos \phi \sin(l - \lambda) \, \delta s_d$$

The measured range can now be seen to be very close to the geometric range

$$\rho_m = (\rho_u + \rho_d)/2 = \rho + 0.5 \, (rR/\rho) \cos \phi \sin(l - \lambda) \, (\delta s_d - \delta s_u)$$

The last term contains the difference between the two angles, which has to be obtained from

$$\delta s_d - \delta s_u = (\psi/c) \, (\rho_d - \rho_u) = (\psi rR/c\rho) \cos \phi \sin(l - \lambda) \, (\delta s_d + \delta s_u)$$

which is of the order of a few times 10^{-11} radians. The result is that we do not need to spend any effort to solve the light-time equation for ranging a geostationary spacecraft, since the directly measured range equals the geometric range in the Earth-fixed system

$$\rho_m = c \, (t_3 - t_1)/2 \approx \rho$$

with the accuracy of better than a millimetre. The only correction needed for the ranging is to time-tag the measurement with the mid-point of the ranging interval, which is very close to t_2. We can use, with an accuracy of 0.1 microseconds during which the spacecraft cannot move a millimetre, the approximation:

$$t_2 \approx (t_1 + t_3)/2$$

The transformation between the Earth-fixed frame and the inertial frame MEGSD shown here is analogous to transformations between frames with mutual rectilinear motion in the Theory of Relativity. Angles are distorted by $V/c = 10^{-5}$ radians and distances by the factor $(V/c)^2 = 10^{-10}$. On the other hand, MEGSD has an almost translational velocity of 30 km/s since it is centred at the Earth which orbits the Sun. This transformation need not be performed here since we use the system MEGSD as reference for all orbital calculations.

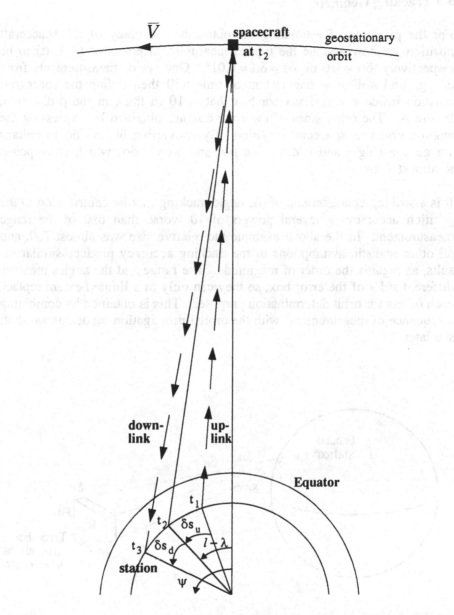

Figure 8.2.C. Schematic drawing, not to scale, of the positions in MEGSD, seen from north, of the ground station and the spacecraft at their respective transmissions and receptions of the ranging signal.

8.3 Tracking Geometry

For the purposes of estimating the obtainable accuracy of the spacecraft position we now assume the typical measurement errors of $(\rho, \varepsilon, \alpha)$ to be respectively, $\delta\rho = 10$ m, $\delta\varepsilon = \delta\alpha = 0.01°$. One set of measurements from one ground station at one instant of time will then define the spacecraft position inside a very flat error box that is 10 m thick in the ρ-direction, Figure A. The other sides of the error box are obtained by expressing the angular errors as spacecraft position. By converting $\delta\varepsilon$ and $\delta\alpha$ to radians we get the height and width to be $\rho\delta\varepsilon$ and $\rho\cos\varepsilon\,\delta\alpha$, which corresponds to almost 7 km.

It is a striking characteristic of the angle tracking that its contribution to the position accuracy is several powers of 10 worse than that of the range measurement. In the above example the relative size was almost 700, and all other realistic assumptions of the tracking accuracy produce similar results, as regards the order of magnitude. The range and the angles measure different sides of the error box, so they can only to a limited extent replace each others for orbit determination purposes. This is obtained by combining a sequence of measurements with the orbital propagation model, as we shall see later.

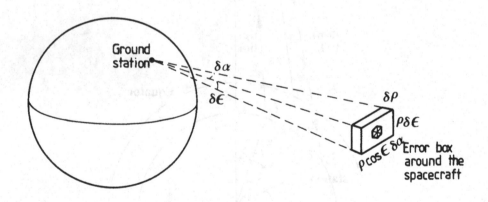

Figure 8.3.A. Perspective view of the flat error box of single station tracking. Not to scale.

The fundamental difference in orbit determination theory between spacecraft in geostationary and those in other types of orbit lies in the very small variations in the ground-station-to-spacecraft geometry of the former. In other orbits, the geometry changes during a station pass and between passes so that one can normally obtain sufficient tracking information for orbit determination after a few passes, even when using only one type of tracking. For geostationary orbits, however, the situation is different in that there are limitations on the information gathered on each orbital element, regardless of the volume of tracking data, since the station always sees the spacecraft in the same direction. The orbit determination process can only use the small variations in spacecraft position that are caused by the deviation of the orbital elements from the geostationary values.

In order to study the measurements of small variations in the spacecraft position we will continue to use the spherical Earth approximation ($f = 0$, $h = 0$). When we vary the spacecraft co-ordinates in the spherical system (r, λ, θ) we obtain a box with sides ($\delta r, r\delta\lambda, r\delta\theta$), Figures B and C, which shall fit inside the station keeping control box in longitude and latitude. Since θ is near zero we omit the factor $\cos\theta$ that would appear in a general variational equation.

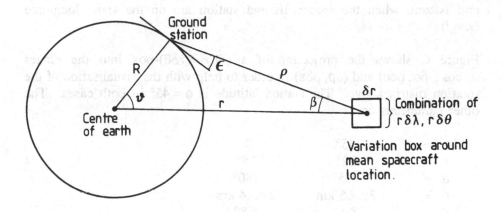

Figure 8.3.B. Schematic view, not to scale, of the plane through the Earth's centre, ground station and spacecraft.

After some lengthy algebraic manipulations we can obtain the following equation for the variation in the tracking parameters $(\rho, \varepsilon, \alpha)$, resulting from the spacecraft position variation, by means of a rotation matrix with two angles:

$$
\begin{pmatrix} \delta\rho \\ \rho\,\delta\varepsilon \\ \rho\,\cos\varepsilon\,\delta\alpha \end{pmatrix} = \begin{pmatrix} \cos\beta & -\sin\beta\,\sin\gamma & -\sin\beta\,\cos\gamma \\ \sin\beta & \cos\beta\,\sin\gamma & \cos\beta\,\cos\gamma \\ 0 & -\cos\gamma & \sin\gamma \end{pmatrix} \begin{pmatrix} \delta r \\ r\,\delta\lambda \\ r\,\delta\theta \end{pmatrix}
$$

Here we have introduced the auxiliary angles β and γ from the relations

$\cos\beta = [r - R\cos(l - \lambda)\cos\phi]/\rho$

$\sin\beta = (R/r)\cos\varepsilon$

$\cos\gamma = r\sin\phi/\rho\cos\varepsilon$

$\sin\gamma = r\sin(l - \lambda)\cos\phi/\rho\cos\varepsilon$

The angle β (Figure B) is the angle between the ground station and the centre of the Earth as seen from the spacecraft. It is always non-zero if we exclude the case $\varepsilon = 90°$, which would have been a station at the sub-satellite point. On the other hand, β is not higher than $8.2°$ as long as we use stations with elevation above $20°$. The angle between the plane of the paper in Figure B and the Earth's axis is denoted by γ. It can take any value and is zero when the spacecraft and station are on the same longitude $(\lambda = l)$.

Figure C shows the projection of a $(\delta r, r\delta\lambda, r\delta\theta)$-box into the planes $(\rho\cos\varepsilon\,\delta\alpha, \rho\delta\varepsilon)$ and $(\delta\rho, \rho\delta\varepsilon)$ in order to help with the visualisation of the rotation matrix above. The station latitude is $\phi = 45°$ in both cases. The other parameters are:

$l - \lambda =$	45°	0°
ε =	22°	38°
α =	235°	180°
ρ =	39365 km	37924 km
β =	8.1°	6.8°
γ =	35.3°	0°

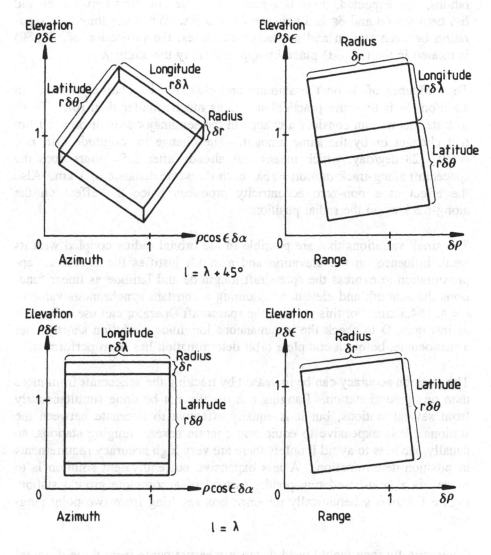

Figure 8.3.C. Variations in azimuth and elevation (left) and in range and elevation (right) from variations in spacecraft position of $\delta r = r\delta\lambda = r\delta\theta = 1$ km (or some other small unit length). The spacecraft and station are on different longitudes (top) and on the same longitude (bottom). Note that the plane of the paper in the right-hand plots is the same as in Figure B.

Examining the rotation matrix or Figure C we notice some qualitative relations. As expected, there is a relatively close coupling between δr and $\delta \rho$, between $\delta \theta$ and $\delta \varepsilon$ and between $\delta \lambda$ and $\delta \alpha$. When the longitude separation between station and spacecraft increases, the projection of $(\delta \lambda, \delta \theta)$ is rotated in the $(\delta \varepsilon, \delta \alpha)$ plane by approximately the angle γ.

The influence of δr on the azimuth and elevation is relatively small. In addition, δr is in many practical situations much smaller than $r \delta \lambda$. To illustrate this we can consider a change in the semimajor axis of $\delta a = \pm 1$ km that changes δr by the same amount. The change in longitude drift rate $= \pm 0.0128$ deg/day, which means that already after 2.54 hours does the spacecraft along-track motion $= r \delta \lambda$ reach the same distance of 1 km. Also the effect of a non-zero eccentricity produces twice the effect on the along-track as on the radial position.

The small variations that are possible in the orbital radius coupled with its weak influence on the elevation and azimuth justifies the use of an approximation to express the spacecraft longitude and latitude as linear functions the azimuth and elevation, assuming a constant synchronous value of $r = 42164.5$ km. For this purpose the spacecraft Operator can use a diagram as in Figure D to check the instantaneous longitude evolution shortly after a manoeuvre, before a complete orbit determination has been performed.

The position accuracy can be increased by tracking the spacecraft from more than one ground station. Ranging can usually not be done simultaneously from several stations, but it is equally effective to alternate between the stations. It is expensive to equip and operate several ranging stations, so usually one tries to avoid it unless there are very high accuracy requirements in position determination. A less expensive but equivalent solution is to range via a land-based transponder instead of a complete ground station. Figure E shows schematically the error box resulting from two-point ranging.

Geometrically, one would need distance measurements from three different stations to determine the instantaneous position of a point in space. However, the ranging requirements can be reduced to *two* stations when several measurements are taken over a time span of a few days and combined with the equation of orbital propagation, but *one* station is not enough, as will be shown later. A combination of range and antenna angle tracking from two or more stations can also be used. If the stations are widely separated, the angles are only of secondary interest because of their relatively lower accuracy as compared with the ranging. They are, however, still useful for enabling a verification of the orbit solution and for improving the confidence in the result, as shown in the next section.

It is easy to understand that the position determination accuracy increases with increasing distance between the two stations used for ranging, provided of course that both stations have visibility above 20° elevation. For geographical and political reasons, however, the stations cannot always be located where it would have been optimal. A further limiting factor can often be the field of view of the spacecraft onboard antenna. In order to estimate the minimum acceptable distance between two stations used for ranging, we can compare the resultant accuracy in longitude or latitude determination with the use of the antenna angles. By simple geometric considerations, assuming the ranging accuracy $\delta\rho = 10$ m and the angle accuracy $\delta\alpha = 0.01°$ converted to radians, we obtain the break-even station distance $2\delta\rho/\delta\alpha = 115$ km = 1° of arc of a great circle on the Earth's surface.

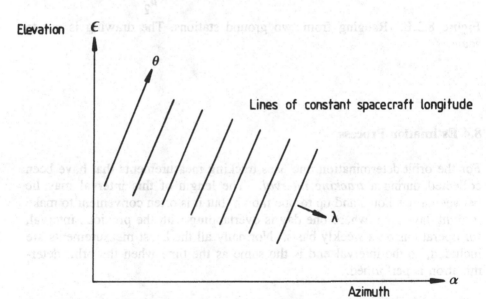

Figure 8.3.D. A diagram of this type can be used by the spacecraft Operator to read off the instantaneous spacecraft longitude from the ground antenna azimuth and elevation.

It is often desirable to temporarily use ranging from one additional station at the beginning of a new mission in order to calibrate the range and antenna angles of a dedicated mission station. Ranging also needs to be calibrated by means of a calibration tower near the station, whereas the angles can be calibrated with a radio star, but this does not give the same confidence in the performance of the system as the use of supplementary stations on a live spacecraft.

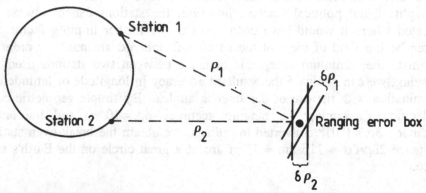

Figure 8.3.E. Ranging from two ground stations. The drawing is not to scale.

8.4 Estimation Process

For the orbit determination one uses tracking measurements that have been collected during a *tracking interval*. The length of this interval may lie between a few hours and up to one month, but it is often convenient to make it eight days long, where one day is overlapping with the previous interval, for operations on a weekly basis. Normally all the latest measurements are included, so the interval end is the same as the time when the orbit determination is performed.

Orbit determination is the process of using the tracking data to calculate a set of parameters, from which one can derive the position and velocity of the spacecraft at any instant inside the tracking interval and also for at least the same time interval into the future. Depending on the character and the purpose of the mission, this can be done in a variety of different ways.

In order to limit the scope of this presentation we will exclude two types of missions that lie outside our present field of interest. First, we will disregard missions for geodetic or similar purposes with accuracy requirements on spacecraft position better than a few tens of metres. Second, we will only consider orbits that are continuously monitored by a control centre and exclude a first acquisition determination method. Apart from these restrictions, the principles described in this section apply to a great extent also to missions other than geostationary ones.

The first one of the above criteria means that we can consider all gravitational attractions on the spacecraft to be exactly known. The Earth is modelled as a solid body, and the details of its motion and of all coordinate systems are assumed to be known. Other perturbations may be only partly known, like the effective cross-section to mass ratio of the solar-radiation pressure, the orbit manoeuvre errors and the perturbations of the attitude control system.

As a result, the spacecraft orbit inside the tracking interval can be described by a small number of parameters that are called *solve-for* parameters and *consider* parameters. The former are estimated in the orbit determination process and will be referred to as p_i, $i = 1,2,...M$ in the following. The latter are parameters that are not sufficiently observable in the current orbit determination and for which one uses values from a preceding determination.

As solve-for parameters one shall take at least the six orbital elements at an epoch, which can be arbitrarily selected inside the interval. Usually the epoch is taken to be the beginning or the end of the tracking interval. As elements one can use the six synchronous elements of Section 2.3 or the six components of the state vector of Section 2.2. It is not possible to use the classical elements, because they become singular and their derivatives discontinuous.

When the tracking interval length is several days without any manoeuvre and there is a sufficient amount of tracking data, one may also include the solar radiation pressure coefficient among the solve-for parameters. If there is one manoeuvre, its $\Delta \overline{V}$ but not the solar coefficient should be estimated. With several manoeuvres one usually solves only one of them. For a long thrust one can solve for the mean acceleration but not the detailed acceleration profile. Only the thrust sizes but not execution times are estimated, since the times should be known with sufficient accuracy beforehand.

If there is redundancy in the tracking data one can also solve for a constant or otherwise systematic bias of the tracking system. Although this is not directly needed for integrating the orbit, one can include it among the solve-for parameters for later use in the tracking equation. With a good orbit determination program one shall be able to select different combinations of solve-for parameters for different situations.

The spacecraft position and velocity at any time inside the tracking interval can now be obtained from a given set of p_i by numerical integration forward or backward from the epoch. We assume that the errors in the orbit are caused only by the errors in the determination of p_i and of remaining errors in the consider parameters. Other errors in the orbital propagation model and from the numerical process are negligible in comparison. This is usually valid for the moderately long integration intervals that we are considering. In the terminology of estimation theory, the orbit determination process described here has no system noise but only measurement noise. For this reason it is suitable to use a *least squares* estimation method and not a numerical filter.

For the estimation process of the parameters p_i, $i = 1,2,...M$ in the orbit determination one needs a sequence of tracking measurements of at least two different types, as will be shown in the subsequent sections. The measurements of all types will here be called q_j, $j = 1,2,...N$ and the corresponding times t_j, $j = 1,2,...N$, where $N > M$.

One can calculate accurately the set of numbers q_j that correspond to a given p_i, but not the reverse because the equations are not linear. It is done by numerical integration of the equations of spacecraft motion, combined with the equations of station visibility of Section 8.1. It is further possible to express the approximate, linearised, dependence of q_j on p_i by combining the equations of Sections 2.3 and 8.3. The final solution is obtained by an iterated linearised least squares fit called the *differential correction* method.

As starting values of p_i in this iteration we use the predicted orbit from the previous determination, the predicted manoeuvre thrusts and the predicted solar radiation pressure coefficient. This explains the above condition with the continuity of the orbit control. The more accurate the prediction is, the faster the iteration converges. If an unplanned orbit perturbation has occurred, e.g. owing to an emergency attitude manoeuvre, the Operator may have to try out different starting orbits to obtain convergence.

We will number the iteration steps l and the corresponding updates of the solve-for parameters by $p_i^{(l)}$. At iteration step number $l + 1$ we integrate the orbit, using the parameters $p_i^{(l)}$, through the tracking interval and interpolate to obtain the simulated spacecraft positions at the times t_j, $j = 1,2,...N$. We can now calculate the *simulated tracking* measurements $q_j^{(l)}$, which would be the distance and angles from the stations to the simulated spacecraft positions at times t_j, $j = 1,2,...N$.

One can describe the simulated tracking as the measurements one would have had if the parameters $p_i^{(l)}$ of the last iteration had been correct and if the tracking system had been free of errors. The differences between the real and the simulated tracking measurements $q_j - q_j^{(l)}$ are called the *tracking residuals* and indicate how close $p_i^{(l)}$ are to the solution. These differences are used to obtain the updated $p_i^{(l+1)}$ at iteration step number $l + 1$ by means of the following least squares fit:

$$p_i^{(l+1)} = p_i^{(l)} + \sum_{j=1}^{N} U_{ij}^{(l)}(q_j - q_j^{(l)}) \quad \text{for } i = 1,2,...M$$

Here $U^{(l)}$ is a matrix with M rows and N columns that expresses the linearised dependence of p_i on q_j. It does not always need to change with the iteration step number (l) but can often be taken as constant.

The orbit determination iteration is considered to have converged when $p_i^{(l)}$ no longer changes more than the expected determination accuracy between one iteration and the next. In many practical cases with well-behaved spacecraft and tracking systems one normally needs only up to five iterations for an orbit determination. After convergence we have approximately

$$\sum_{j=1}^{N} U_{ij}^{(l)}(q_j - q_j^{(l)}) = 0 \quad \text{for } i = 1,2,...M$$

but this does not necessarily imply that the final tracking residuals $q_j - q_j^{(l)} = 0$ for all $j = 1,2,...N$. It only proves that the tracking data is *consistent*. Consistent tracking means that there is no contradiction between the available tracking measurements and the determined orbit, but it does not always mean that the latter is very accurate.

Consistency can also be caused by too little tracking data or by determination of too many solve-for parameters. One can only know that consistency implies a good result if one has calculated, by theoretical means, that the determined parameters have good observability with the given tracking configuration. There should also be reasonable agreement between the determined orbit and the orbit predicted from the previous determination to ensure confidence in the result.

If the tracking residuals are randomly distributed in j we know that the errors are caused by random noise and the orbit determination accuracy has reached its ultimate limit. If there is some trend in the residuals we may suspect that there is a bias in the tracking system or an additional perturbation in the orbital propagation model. A typical example of such a trend is a residual that is linear in time and passes through zero at the centre of the tracking interval. Another type may be periodic with a period of one day and a constant or linearly changing amplitude.

If one discovers one of these trends one can improve the orbit determination accuracy by repeating it with one more solve-for parameter, which must be chosen according to the trend. This requires, of course, that there is so much tracking data that one can distinguish, say, a randomly and a periodically distributed residual. On the other hand one must not introduce more solve-for parameters than are observable by the tracking data. For these reasons, the operation of an orbit determination process is more an art than a craft, which requires both theoretical insight and practical experience on the part of the Operator.

The differential correction method described here for orbit determination has a certain similarity to the mathematical method in Section 7.6 for calculating manoeuvres. In both cases there is an accurate model that is combined in an iteration with the inverse of an approximate model. There are, however, some important differences. In Section 7.6, the number of variables to be solved is the same as the amount of input, and the result is, up to a point, independent of the approximate model. Further, the input data consists of exactly defined target parameters. On the other hand, the orbit determination estimates a few parameters from a larger number of physical measurements that are exposed to unmodelled and partly unknown errors, and the result depends on the selection of the approximate model.

It is easy to realise from the last equation that, because $N > M$, one will get a different set of solutions $q_j^{(l)}$ for different values of the matrix U, which is derived from the approximate model. For this reason one ought to select the matrix with great care, but in practice this is seldom possible, as is shown in the following. It is general practice to compose it from the *weighting matrix* W, the matrix of *partial derivatives* Q and its transpose Q^T as follows:

$$U = (Q^T W Q)^{-1} Q^T W \quad \text{with} \quad Q_{ji} = \frac{\partial q_j}{\partial p_i}.$$

The matrix W is used to give a higher weight to the more accurate types of tracking and lower to the less accurate types. In particular, its purpose is to weight different types of measurements, e.g. angles versus distances, into the same least squares fit. An important property of W is that it must be a positive definite matrix. According to the classical *Gauss-Markoff* theorem, the optimum value of W that minimises the variance of the remaining errors is the inverse of the covariance matrix of the measurement errors.

The reason why one can rarely obtain an optimal solution is that in most practical cases the exact statistical properties of the tracking errors are unknown, and only approximate assumptions are available. Often W is taken to be diagonal with the elements inversely proportional to the square of what one would expect the errors of the corresponding tracking types to be. A further approximation in the calculation of the ideal matrix U is due to approximations of the partial derivatives in Q. However, since it only appears in the formulas in combination with W one does not gain any accuracy by calculating it with a higher relative precision than with what W is known. Note that the permissible relative error in Q shall be compared with, not the tracking errors that constitute W, but the relative *error in the knowledge of the errors* of the tracking measurements, which normally is much higher.

Approximate formulas for the partial derivatives are given in the following sections. The matrix element Q_{ji} expresses by how much one expects the tracking measurement q_j to change if the solve-for parameter p_i is varied by a small amount. Of course one can only solve a p_i if its value has any measurable influence on at least one member of q_j. One then says that the parameter p_i is *observable* with the given tracking data. It may happen that changes of two different p_i have the same influence on the set of q_j and then it is not possible to tell which p_i is responsible. We will see examples of this situation later, where a subset of p_i has poor observability.

If there is bad observability of one or a subset of the parameters p_i, then the matrix inversion

$$(Q^T W Q)^{-1}$$

becomes singular and U cannot be calculated. There are also varying degrees of observability that give more or less well-conditioned matrix inversions. A badly conditioned inversion results in a U-matrix with some elements that are abnormally high in absolute values. This usually causes the iteration to diverge instead of converge, apart from introducing other numerical difficulties in the computation.

If the amount of tracking data available is very small, the data may not be enough for determining all the six orbital elements. In some situations one may temporarily have to determine only the four in-plane orbital elements and use predicted values for the two out-of-plane elements (i_x, i_y) or ($z, dz/dt$) at the epoch. Such a reduced orbit determination may be used for an in-orbit spare spacecraft, which is only rarely manoeuvred and has modest requirements on orbit accuracy and must economise on the amount of tracking data. Another situation arises when one has only a short tracking interval of, e.g. a few hours, and needs to determine quickly the longitude and its drift rate after an unplanned orbit disturbance, as described in Section 7.5.

8.5 Derivatives and Differentials

For calculating the matrix Q of approximate partial derivatives in the orbit determination formula of the preceding section one can use the already derived linearised relations for unperturbed orbital motion of Section 2.3 and for the tracking geometry of Section 8.3. The tracking measurements q_j consist of a set of values of ($\rho, \varepsilon, \alpha$) at different times t_j and possibly taken from different ground stations.

For the derivation of the partial derivatives we will here make use of the concept of *differentials* from the theory of differential calculus. Differentials can be visualised as small variations of the parameters, in this case δq_j of the tracking measurement q_j and δp_i of the solve-for parameter p_i.

The partial derivative matrix Q shows how much the tracking measurement would have been changed by a small change in the solve-for parameter:

$$\delta q_j = \sum_{i=1}^{M} Q_{ji}\, \delta p_i \quad \text{for } j = 1,...N$$

When p_i is a tracking bias it is easy to calculate the element Q_{ji} that is needed to determine it. It is equal to +1 or 0 depending on if the corresponding q_j is measured by that tracking type or not. Normally the 6 orbital elements shall be solved with the solve-for parameters

$$(p_1,\ p_2,\ p_3,\ p_4,\ p_5,\ p_6) = (\lambda_0,\ D,\ e_x,\ e_y,\ i_x,\ i_y)$$

The linear dependence of the element differentials ($\delta\lambda_0$, δD, δe_x, δe_y, δi_x, δi_y) and the tracking differentials ($\delta\rho$, $\delta\varepsilon$, $\delta\alpha$) is calculated via the spacecraft position (δr, $\delta\lambda$, $\delta\theta$) at the times of tracking t_j. We start with the approximate orbital motion of Section 2.3 that provides the spacecraft position at time t, converted to the spacecraft sidereal angle by:

$$s = s_0 + \psi(t - t_0)$$

$$r = A(1 - 2D/3 - e_x \cos s - e_y \sin s)$$

$$\lambda = \lambda_0 + D(s - s_0) + 2e_x \sin s - 2e_y \cos s$$

$$\theta = -i_x \cos s - i_y \sin s$$

According to Section 4.1 the natural perturbations contribute to this motion in the first order by additional terms that do not depend on the orbital elements. As an example, the quadratic term that provides the parabolic time-dependence of the mean longitude plotted in Figure 4.2.B has the coefficient

$$\ddot{\lambda} = -3B/A$$

which is well known and constant as long as the spacecraft is confined to a narrow deadband. Such terms then vanish when one takes the derivatives with respect to the orbital elements, so the unperturbed motion provides a good approximation for the partial derivatives.

By defining the matrix Φ, which is a function of s, one can write the equation for the differentials:

$$\Phi(s) = A \begin{pmatrix} 0 & -2/3 & -\cos s & -\sin s & 0 & 0 \\ 1 & s-s_0 & 2\sin s & -2\cos s & 0 & 0 \\ 0 & 0 & 0 & 0 & -\cos s & -\sin s \end{pmatrix}$$

$$\begin{pmatrix} \delta r \\ A\delta\lambda \\ A\delta\theta \end{pmatrix} = \Phi(s) \begin{pmatrix} \delta\lambda_0 \\ \delta D \\ \delta e_x \\ \delta e_y \\ \delta i_x \\ \delta i_y \end{pmatrix}$$

We introduce now the matrix Ψ to represent the spacecraft-to-station geometry for the relation between the tracking differentials and the spacecraft position that was derived in Section 8.3. The matrix is a function of the ground station latitude ϕ and longitude relative to the spacecraft, $(l-\lambda)$, from which ρ and ε as well as the auxiliary angles β and γ can be derived. We insert $r = A$ for the geostationary radius.

$$\Psi = \begin{pmatrix} 1 & 0 & 0 \\ 0 & \rho & 0 \\ 0 & 0 & \rho\cos\varepsilon \end{pmatrix}^{-1} \begin{pmatrix} \cos\beta & -\sin\beta\sin\gamma & -\sin\beta\cos\gamma \\ \sin\beta & \cos\beta\sin\gamma & \cos\beta\cos\gamma \\ 0 & -\cos\gamma & \sin\gamma \end{pmatrix}$$

$$\begin{pmatrix} \delta\rho \\ \delta\varepsilon \\ \delta\alpha \end{pmatrix} = \Psi \begin{pmatrix} \delta r \\ A\,\delta\lambda \\ A\,\delta\theta \end{pmatrix}$$

One can now obtain the complete matrix of partial derivatives Q from the two matrix equation above by equating the differential of the spacecraft position $(\delta r, A\delta\lambda, A\delta\theta)$ that appears in both. The equation for an element Q_{ji} of the matrix of partial derivatives becomes:

$$Q_{ji} = \sum_{k=1}^{3} \Psi_{qk}(j)\, \Phi_{ki}(s_j) \quad \text{for} \quad i = 1,...6, \quad j = 1,...N$$

Since different ground stations may be used for the different measurements we have denoted the elements of Ψ to be functions of the index j. In most practical cases there are only one or two stations that take a large number of measurements each. The index $q = 1, 2$ or 3 when the tracking measures, respectively, the range, the elevation or the azimuth. Measurements are labelled with the index j and are taken at the times t_j when the sidereal angle of the nominal spacecraft position is:

$$s_j = s_0 + \psi(t_j - t_0)$$

In the same way as before we can obtain the differentials of the spacecraft position from a manoeuvre thrust $\Delta \overline{V}$ executed at the time t_b. We denote the differential of $\Delta \overline{V}$ by $\delta \Delta \overline{V}$ and its radial, tangential and orthogonal components by the indices r, t and o, respectively. From the formulas in Sections 3.2 and 3.3 we derive, with the notation

$$\Delta s = s - s_b = \psi(t - t_b)$$

$$
\begin{pmatrix} \delta r \\ A\delta\lambda \\ A\delta\theta \end{pmatrix} = \frac{1}{\psi} \begin{pmatrix} \sin \Delta s & 2 - 2\cos \Delta s & 0 \\ 2\cos \Delta s - 2 & 4\sin \Delta s - 3\Delta s & 0 \\ 0 & 0 & \sin \Delta s \end{pmatrix} \begin{pmatrix} \delta\Delta V_r \\ \delta\Delta V_t \\ \delta\Delta V_o \end{pmatrix}
$$

This formula is valid only for the tracking at times t_j that lie on the opposite side of t_b, counted from the epoch at which the elements are defined. For the tracking at times on the same side of t_b as the epoch, the influence of the thrust is of course zero. The matrix in the equation above is then used for $\Phi(s)$ in the previous formula to calculate Q as before. The effect of several manoeuvres in the tracking interval is handled by superposition of the influence from each single manoeuvre. On the other hand, it is not easy to include the thrust time t_b among the solve-for parameters, but this is practically never required if the manoeuvres are executed by ground commands at well-known times.

For the effective cross-section to mass ratio σ of the solar radiation pressure we use the mean influence during one day, given in Section 4.5. We can consider, approximately, the right ascension s_S of the Sun to be constant during the tracking interval. When t_0 is the epoch, one obtains the spacecraft position differential to be:

$$
\begin{pmatrix} \delta r \\ A\delta\lambda \\ A\delta\theta \end{pmatrix} = \frac{3P(t - t_0)}{2\psi} \begin{pmatrix} -\sin(s - s_S) \\ -2\cos(s - s_S) \\ 0 \end{pmatrix} \delta\sigma
$$

These formulas for the partial derivatives can now be used in a computer program for orbit determination, where all the calculations are performed numerically. The analytical methods that will be derived in the following are not needed in the program, but are intended to give the reader an insight into the theory of the possibilities and limitations in orbit determination and contribute to the understanding of how the parameter observability depends on the tracking data. We start by studying the variations, as functions of time, of the spacecraft position differentials and the tracking differentials.

In general, each of the three components of the unperturbed spacecraft motion shown in Figure 2.3.C can be described, as functions of the sidereal angle s, by: A sum of four terms that are either constant, linear or periodic as cosine or sine. Because each tracking differential is a linear combination of the three components of the spacecraft position, they all have the same dependence on time and the sidereal angle s, as plotted in Figure A:

$$\delta q = b + u\,(s - s_0) + v \cos s + w \sin s$$

Here δq stands for a tracking differential ($\delta\rho$, $\delta\varepsilon$, $\delta\alpha$). The 4 coefficients (b,u,v,w) are independent of time and the sidereal angle but depend on the type of tracking and on the location of the ground station. Note that this time dependence only is valid for δq and not for q itself (i.e. ρ, ε, α), since the complete expression for q also contains the short- and long-periodic natural perturbations. However, for approximate calculations one can sometimes use a similar time-dependence also for the latter.

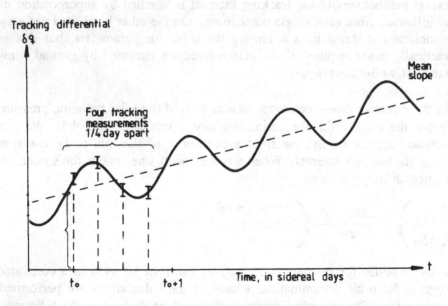

Figure 8.5.A. The time dependence of a tracking differential can be expressed as the sum of a constant, a linear, a cosine and a sine term. The four coefficients are calculated from at least four tracking measurements.

It is obvious from Figure A that at least 4 measurements are needed of one tracking type to determine its 4 coefficients (b,u,v,w). Their time distribution is important for the determination accuracy. The measurement times must not be separated by an integral multiple of 12 hours, because then v and w can not be determined. The linear coefficient u is best determined from widely separated data points. A good distribution of points is to track every 4 or 6 hours during the whole tracking interval, which should be typically 8 days long. If one is allowed only a small number of measurements but has the choice of how to distribute them over a long tracking interval, then it is not optimal to spread them evenly. The preferred solution is to take at least 4 points, 6 hours apart, at the beginning of the interval and another 4 points at the end of the interval.

When there is a single manoeuvre inside the tracking interval, the tracking differentials have the same type of time dependence as above on either side of the thrust but with different sets of coefficients (b,u,v,w). The coefficients are, however, not completely independent, since the curves must meet at the time of the thrust, Figure B. Instead of describing this time dependence by eight coefficients (b,u,v,w), four on each side of the thrust, one only needs seven parameters, which can be determined from at least seven tracking measurements. This reduction in the number of solve-for parameters is the reason why it is preferable to determine the size of manoeuvre thrusts inside the tracking interval instead of at the boundary between two tracking intervals, as mentioned in Section 5.2. The determined ΔV can then be used for calibrating the thrust as described in Section 3.1.

When there are several manoeuvre thrusts inside the tracking interval, they must be separated by at least about half a day and three tracking measurements to make is possible to solve each size, Figure C. If there is not enough tracking between the thrusts one can instead consider all $\Delta \overline{V}$s except one to be known and only solve for the three components of this remaining $\Delta \overline{V}$. The combined error of all the thrusts in the tracking interval will then be allocated to only one of them.

Another method to reduce the number of solve-for parameters after one or several longitude manoeuvres when there is little tracking data, is to solve for only one component of each $\Delta \overline{V}$, namely the tangential component. For this purpose it would in theory be sufficient to have one tracking measurement between each manoeuvre and after the last one. However, this is only

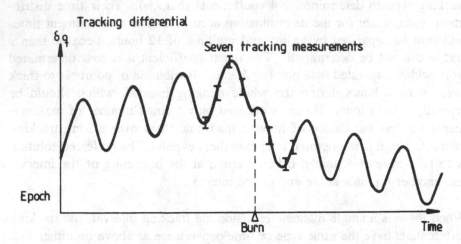

Figure 8.5.B. Time-dependence of a tracking differential when a thrust $\Delta\overline{V}$ is included among the solve-for parameters. Seven coefficients can be determined from seven tracking measurements, four before and three after the thrust.

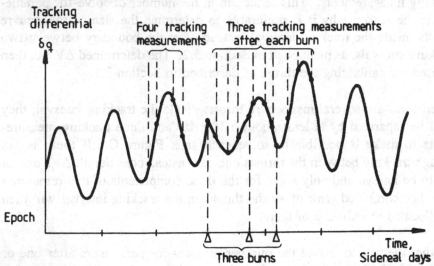

Figure 8.5.C. In order to solve three components of each thrust ΔV one needs at least three tracking measurements and half a sidereal day after each thrust.

recommended in non-nominal situations, e.g. when a quick result is needed for checking the longitude drift rate, as in Section 7.5.

For the determination of only the tangential component of one manoeuvre $\Delta \overline{V}$ we make use of the already given expressions for the longitude differential $\delta \lambda$. Its evolution $\delta \lambda_1$ as it would have been without the thrust is plotted in Figure D according to:

$$\delta \lambda_1 = \delta \lambda_0 + (s - s_0) \, \delta D + 2 \sin s \, \delta e_x - 2 \cos s \, \delta e_y$$

For the time on the other side of the thrust there is an additional term that is proportional to ΔV_t:

$$\delta \lambda_2 = \delta \lambda_1 + [4 \sin(s - s_b) - 3(s - s_b)] \, \Delta V_t / V$$

The orbit determination is in principle performed by solving the four elements (λ_0, D, e_x, e_y) from the four tracking measurements taken before the thrust, Figure D. By the use of these elements the evolution of $\delta \lambda_1$ can be calculated also for the time after the thrust. By inserting this $\delta \lambda_1$ into the last equation one is able to calculate ΔV_t by performing one single tracking measurement that gives a measure of $\delta \lambda_2$ at one instant because the coefficient in front of ΔV_t is known.

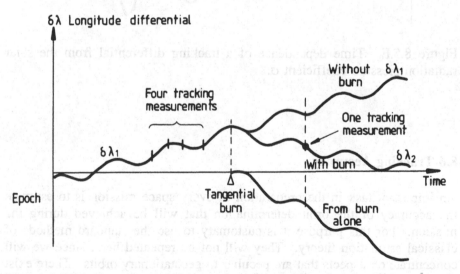

Figure 8.5.D. One measurement point after a thrust is enough to solve the tangential component of the $\Delta \overline{V}$.

Finally, we have the differential to be used for the determination of the solar radiation pressure coefficient σ in front of of the effective cross-section to mass ratio. It has the following general form, where C_1 and C_2 are two known coefficients that can be calculated from the matrix Ψ.

$$\delta q = [C_1 \cos(s - s_S) + C_2 \sin(s - s_S)](t - t_0)$$

This is a sinusoidal oscillation with an approximately linearly changing amplitude. Since this amplitude change is rather weak one can in general only solve σ using relatively long tracking intervals without any orbit manoeuvres, Figure E.

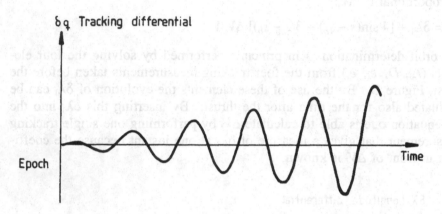

Figure 8.5.E. Time dependence of a tracking differential from the solar radiation pressure coefficient σ.

8.6 Tracking Accuracy

An important task in the preparation of every space mission is to estimate the accuracy of the orbit determination that will be achieved during the mission. For this purpose it is customary to use the standard methods of classical estimation theory. They will not be repeated here, since we will concentrate on aspects that are peculiar to geostationary orbits. There exist some unique analytical relations that are caused by the fact that the position of a geostationary spacecraft relative to the ground stations remains practically constant during the mission.

A general problem in simulations of an expected determination accuracy is
the assumptions to make for the characteristics of the measurement errors.
The frequently made assumption of the error as the sum of a constant bias
and a noise with Gaussian distribution tends to underestimate the determination error unless the results are interpreted carefully. In practice it is
usually found that the error with the highest influence on the result is neither
a constant nor a random noise. Instead, it varies on an intermediate time
scale that is shorter than the tracking interval but longer than the time between the individual tracking measurements.

In most cases only the order of magnitude of the error is known but not the
details of its probability distribution. In general one can state that the simulation result is reliable when it is not strongly dependent on the assumed
statistical properties of the errors. For this reason we will here only briefly
deal with accuracy results that depend on error statistics. The main task is
to obtain deterministic analytical relations between the errors in the orbit
and in the tracking.

The observability of other parameters than the 6 elements will no longer
be dealt with explicitly since they can indirectly be derived from changes
in the elements. For example, the solar radiation pressure coefficient can
be obtained from slow changes of the eccentricity vector. The three components of a manoeuvre $\Delta \overline{V}$, the radial, tangential and orthogonal components, have the same observability as, respectively, the eccentricity vector,
the longitude drift rate and the inclination vector.

The analytical method to estimate the orbit determination accuracy relies
upon the differentials of the orbital elements and of the tracking measurements that were introduced in Section 8.5. The observability and the size
of the errors of the orbital elements is the same as of their respective differentials, because the latter can be considered as the update of the last iteration in the determination with the differential correction method of
Section 8.4.

In Section 8.5 we defined the set of 4 coefficients (b,u,v,w) that are individual for each type of tracking (range, elevation and azimuth) from each
station. They describe the variation of a tracking differential with time,
expressed via the spacecraft nominal sidereal angle s

$$\delta q = b + u\,(s - s_0) + v \cos s + w \sin s$$

The constant s_0 is the value of s at the defined epoch t_0. Here q stands for one type of tracking (ρ, ε, α) from one ground station. The new mathematical method that enables the analytical estimate of the accuracy is based upon a split of the determination process into two parts. This is not used for practical orbit determination, but in the analytical method it provides a good insight into the observability and accuracy. The two parts are:

- Determine the 4 coefficients (b,u,v,w) separately for each type of tracking from each station.

- Determine the 6 orbital elements (λ_0, D, e_x, e_y, i_x, i_y) by combining the 4 coefficients for several tracking types and/or stations.

We now assume that we have got a set of tracking measurements q_j. The index j here no longer refers to all the N measurements, as in Sections 8.4 and 8.5, but only the subset that is taken with one tracking type and from one station. At the measurement times t_j, the spacecraft sidereal angle equals s_j. The tracking differential varies with this angle according to:

$$\delta q_j = b + u\,(s_j - s_0) + v\cos s_j + w\sin s_j$$

We can solve the 4 coefficients (b,u,v,w) from the above tracking equation if there are at least 4 measurement points q_j. When there are exactly 4 points, there is a deterministic solution as illustrated in Figure 8.5.A. With the tracking points at the sidereal angles (with $s_0 = 0$)

$$s_1 = 0 \; ; \quad s_2 = \pi/2 \; ; \quad s_3 = \pi \; ; \quad s_4 = 3\pi/2$$

the solution becomes

$$\begin{pmatrix} b \\ u \\ v \\ w \end{pmatrix} = \begin{pmatrix} 1 & -0.5 & 1 & -0.5 \\ -1/\pi & 1/\pi & -1/\pi & 1/\pi \\ 0 & 0.5 & -1 & 0.5 \\ -0.5 & 1 & -0.5 & 0 \end{pmatrix} \begin{pmatrix} \delta q_1 \\ \delta q_2 \\ \delta q_3 \\ \delta q_4 \end{pmatrix}$$

In this case the error in each one of the calculated (b,u,v,w) is about the same as the mean tracking measurement error in δq_j. It is easy to see that the following qualitative properties are valid, and that they can be generalised to other distributions of measurement points. A constant bias error will affect only b, whereas v and w will be particularly influenced by errors that are variable with a one day period. The latter could be caused by, e.g. the day/night heating and cooling of the antenna structure or of the part of the atmosphere through which the tracking signal passes.

In most cases there are many more than 4 tracking points of each type. This results in an overdetermined system for solving (b,u,v,w) by least squares in the same way as in the complete orbit determination of Section 8.4. Since each fit processes measurements of the same type, it is reasonable to give them the same weight. We introduce the 4×4 matrix Ξ that here fulfils the same function as $(Q^T W Q)^{-1}$ in the complete solution of Section 8.4. Its inverse is defined by:

$$\Xi^{-1} = \sum_j \begin{pmatrix} 1 \\ s_j - s_0 \\ \cos s_j \\ \sin s_j \end{pmatrix} (1 , \ s_j - s_0 , \ \cos s_j , \ \sin s_j)$$

The least squares solution becomes, when there are more than 4 tracking points:

$$\begin{pmatrix} b \\ u \\ v \\ w \end{pmatrix} = \Xi \sum_j \begin{pmatrix} 1 \\ s_j - s_0 \\ \cos s_j \\ \sin s_j \end{pmatrix} \delta q_j$$

A tracking campaign that was reported in the ESA Journal volume 10 of 1986, pages 71 to 83, measured directly the errors in the angles of the 13.5 m 11 GHz TMS-1 antenna at Redu (Belgium) with the help of ESA's communications spacecraft ECS-1 and OTS-2. For the evaluation, a highly accurate orbit was determined by two station ranging from Redu and Villafranca (Spain). With this orbit, the simulated elevation and azimuth angles were calculated and subtracted from those measured. The resulting error was used in the formula above with δq = either $\delta \varepsilon$ or $\delta \alpha$ to calculate the corresponding set (b,u,v,w).

There were 6 angular measurements made per day, and the tracking intervals were 2, 4 or 8 days long. About 200, partly overlapping, intervals were processed. The most important result is the statistics of the amplitude error

$$\sqrt{v^2 + w^2}$$

Its standard deviation will in the following be denoted by σ_{ampl}. The distribution function is plotted in Figures A, B, C and D. The vertical axis shows, in percent, the number of cases found with an error smaller than the value on the horizontal axis. As expected, the azimuth is more accurate than the elevation, since the refraction in the atmosphere is highest for the latter, Section 8.2. The 99% probability level is seen to be:

- Elevation: 0.0045° (8 days) and 0.007° (2 days)

- Azimuth: 0.0025° (8 days) and 0.005° (2 days)

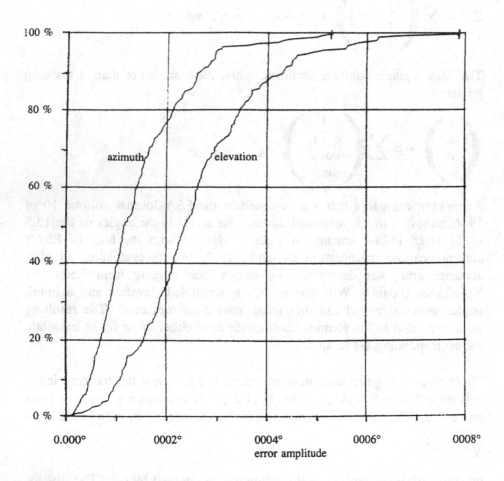

Figure 8.6.A. Distribution function for azimuth and elevation error amplitude of 232 intervals of 2 days, for ECS-1.

We will refer to the 99% level as the $3\sigma_{vw}$ value in the following, which holds true for Gaussian error distributions. In the same campaign, the values of b were found to be of about the same size, up to 0.005°, as the other coefficients and varied between different tracking intervals. The main cause was probably an error of the previously mentioned kind that is neither constant nor completely random.

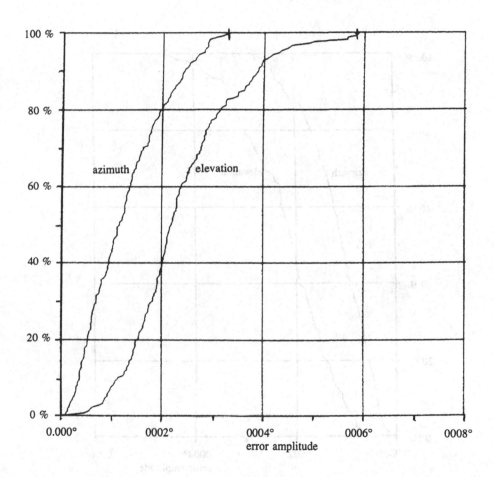

Figure 8.6.B. Distribution function for azimuth and elevation error amplitude of 204 intervals of 4 days, for ECS-1.

The direct measurement of the angular error was possible because it could be compared with separate, more accurate, ranging results. On the other hand, it is difficult to find a more accurate tracking for measuring the ranging error, so there are at present no such detailed error statistics for ranging. In its absence one has to rely upon crude simulations, where only the order of magnitude of the tracking error is known. A common approach is to assume that the errors in δq_j are unbiased and uncorrelated and have the standard deviation $= \sigma_q$. This assumption leads to the following covariance matrix for (b,u,v,w):

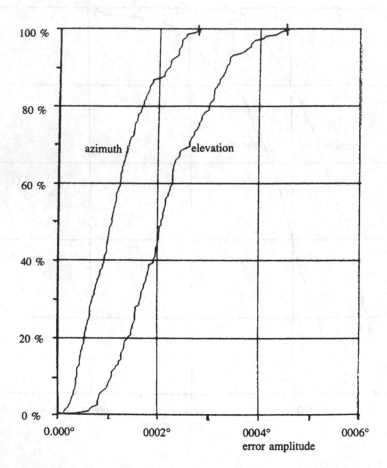

Figure 8.6.C. Distribution function for azimuth and elevation error amplitude of 184 intervals of 8 days, for ECS-1.

$$\begin{pmatrix} cov(b,b) & cov(u,b) & cov(v,b) & cov(w,b) \\ cov(b,u) & cov(u,u) & cov(v,u) & cov(w,u) \\ cov(b,v) & cov(u,v) & cov(v,v) & cov(w,v) \\ cov(b,w) & cov(u,w) & cov(v,w) & cov(w,w) \end{pmatrix} = \Xi \, \sigma_q^2$$

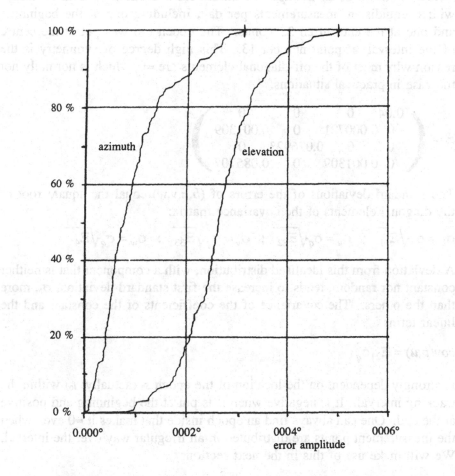

Figure 8.6.D. Distribution function for azimuth and elevation error amplitude of 199 intervals of 8 days, for OTS-2.

The matrix Ξ is calculated according to the previous formula from the sequence of nominal spacecraft sidereal angles s_j at the tracking points. The errors (b,u,v,w) obtained in this way can be inserted into the formulas of Section 8.7 for estimating the expected orbit determination accuracy. Some care must be exercised in such a simulation by not taking too many tracking points. Otherwise the covariance matrix may become unrealistically small.

Below is shown as an example the matrix for a tracking interval of 4 days with 6 equidistant measurements per day, including one at the beginning and one at the end; in all 25 points. The epoch (s_0) was put at the centre of the interval, at point number 13. This high degree of symmetry is the reason why most of the off-diagonal elements are =0, which is normally not the case in practical situations.

$$\Xi = \begin{pmatrix} 0.04 & 0 & 0 & 0 \\ 0 & 0.000721 & 0 & 0.001309 \\ 0 & 0 & 0.076923 & 0 \\ 0 & 0.001309 & 0 & 0.085707 \end{pmatrix}$$

The standard deviations of the errors of (b,u,v,w) equal the square root of the diagonal elements of the covariance matrix:

$$\sigma_b = \sigma_q\sqrt{\Xi_{11}} \;\; ; \;\; \sigma_u = \sigma_q\sqrt{\Xi_{22}} \;\; ; \;\; \sigma_v = \sigma_q\sqrt{\Xi_{33}} \;\; ; \;\; \sigma_w = \sigma_q\sqrt{\Xi_{44}}$$

A deviation from this idealised distribution, with a component that is neither constant nor random, tends to increase the first standard deviation, σ_b, more than the others. The covariance of the coefficients of the constant and the linear terms

$$cov(b,u) = \Xi_{21}\sigma_q^2$$

is strongly dependent on the location of the epoch s_0 (actually: t_0) within the tracking interval. It is negative when it is put at the beginning and positive at the end. One can always find an epoch inside that makes it =0 even when the measurement points are distributed in an irregular way over the interval. We will make use of this in the next section.

Subsequent numerical examples will be based upon the following typical order of magnitude of the error of range in the S-band frequency or higher. It is not only the error of the measurement itself, but includes also the effect of model errors in a normal orbit propagation and tracking software. Experience from many years of orbit determinations with least squares fit has shown this to be the typical ranging residual.

- Range: $\sigma_b = \sigma_{vw} = 5$ m; 99% level: $3\sigma_b = 3\sigma_{vw} = 15$ m

The next step is now to obtain the 6 orbital elements, here represented by their differentials δp_i from several sets of the 4 coefficients. In order to to identify the type of tracking we now attach the index q to the 4 coefficients (b,u,v,w), but we still consider only one ground station. The least squares solution for the 4 coefficients is, from before:

$$
\begin{pmatrix} b_q \\ u_q \\ v_q \\ w_q \end{pmatrix} = \Xi \sum_j \begin{pmatrix} 1 \\ s_j - s_0 \\ \cos s_j \\ \sin s_j \end{pmatrix} \delta q_j
$$

On the right hand side above we insert the expression for δq_j of the complete tracking equation in Section 8.5.

$$
\delta q_j = \sum_{i=1}^{6} Q_{ji}\,\delta p_i = \sum_{k=1}^{3} \Psi_{qk} \sum_{i=1}^{6} \Phi_{ki}(s_j)\delta p_i
$$

The index $q = 1$, 2 or 3 for a measurement of the range, elevation or azimuth, respectively. The result becomes:

$$
\begin{pmatrix} b_q \\ u_q \\ v_q \\ w_q \end{pmatrix} = \sum_{k=1}^{3} \Psi_{qk} \sum_{i=1}^{6} \delta p_i\, \Xi_q \sum_j \begin{pmatrix} 1 \\ s_j - s_0 \\ \cos s_j \\ \sin s_j \end{pmatrix} \Phi_{ki}(s_j)
$$

As before in this section, each sum over j shall only include the subset taken with the applicable type of tracking q. The set of tracking points may be different for different types, which is the reason why Ξ_q depends on q in the equation above. One can see from the definition of Φ that each sum of the following type that appears above

$$
\sum_j \begin{pmatrix} 1 \\ s_j - s_0 \\ \cos s_j \\ \sin s_j \end{pmatrix} \Phi_{ki}(s_j)
$$

also appears in the definition of some element of Ξ_q^{-1}. At the subsequent multiplication by Ξ_q all dependence on individual points s_j is eliminated. The result leads to the equations below, after some algebraic work.

The part that is obtained from the cosine and sine terms is simplified by the use of complex numbers, where we denote the imaginary unit by j ($j^2 = -1$) so as not to confuse it with the inclination. We insert the differentials of the 6 synchronous orbital elements:

$$(\delta p_1, \delta p_2, ... \delta p_6) = (\delta\lambda_0, \delta D, \delta e_x, \delta e_y, \delta i_x, \delta i_y)$$

When $q = \rho = 1$ we obtain:

$$\begin{pmatrix} b_\rho \\ u_\rho \end{pmatrix} = A \begin{pmatrix} -\sin\beta\sin\gamma & -(2/3)\cos\beta \\ 0 & -\sin\beta\sin\gamma \end{pmatrix} \begin{pmatrix} \delta\lambda_0 \\ \delta D \end{pmatrix}$$

$$v_\rho + jw_\rho =$$

$$= -A(\cos\beta + 2j\sin\beta\sin\gamma)(\delta e_x + j\delta e_y) + A\sin\beta\cos\gamma(\delta i_x + j\delta i_y)$$

When $q = \varepsilon = 2$ we obtain:

$$\rho \begin{pmatrix} b_\varepsilon \\ u_\varepsilon \end{pmatrix} = A \begin{pmatrix} \cos\beta\sin\gamma & -(2/3)\sin\beta \\ 0 & \cos\beta\sin\gamma \end{pmatrix} \begin{pmatrix} \delta\lambda_0 \\ \delta D \end{pmatrix}$$

$$\rho(v_\varepsilon + jw_\varepsilon) =$$

$$= A(-\sin\beta + 2j\cos\beta\sin\gamma)(\delta e_x + j\delta e_y) - A\cos\beta\cos\gamma(\delta i_x + j\delta i_y)$$

When $q = \alpha = 3$ we obtain:

$$\rho\cos\varepsilon \begin{pmatrix} b_\alpha \\ u_\alpha \end{pmatrix} = -A\cos\gamma \begin{pmatrix} \delta\lambda_0 \\ \delta D \end{pmatrix}$$

$$\rho\cos\varepsilon(v_\alpha + jw_\alpha) = -2Aj\cos\gamma(\delta e_x + j\delta e_y) - A\sin\gamma(\delta i_x + j\delta i_y)$$

In the next section, the same equations will be derived in a slightly different way and used to find analytical relations between the tracking accuracy and the orbit determination accuracy, as well as the observability of individual orbital elements.

8.7 Observability and Accuracy

The analytical method for estimating the orbit errors in this section relies strongly on the orbit propagation and its agreement with the model in the orbit determination software. For this reason it is only valid to a certain limit in accuracy caused by unavoidable approximations in the modelling of the orbit propagation and in the tracking process. Depending on the spacecraft design, there may be additional, unmodelled, perturbations from the on-board attitude control system, the effect of which on the orbit accuracy increases with decreasing redundancy in the tracking data.

A further limitation to this method is that it does not model a quick orbit determination with only a few tracking measurements of different types, e.g. shortly after an orbit manoeuvre. Apart from these constraints, the presented analytical method can replace a full-scale simulation for estimating the expected determination accuracy of a future mission.

In Section 8.5 we found that the differentials of the tracking measurements can be expressed by the matrix equation

$$
\begin{pmatrix} \delta\rho \\ \delta\varepsilon \\ \delta\alpha \end{pmatrix} = \Psi\,\Phi(s) \begin{pmatrix} \delta\lambda_0 \\ \delta D \\ \delta e_x \\ \delta e_y \\ \delta i_x \\ \delta i_y \end{pmatrix}
$$

The elements of the matrix $\Phi(s)$ are either constant or depend on the spacecraft sidereal angle s as $(s - s_0)$, $\cos s$ or $\sin s$. Each component of the vector on the right hand side $\delta\rho$, $\delta\varepsilon$ and $\delta\alpha$ must then be linear combinations of these functions. For this dependence on s we have introduced the 4 coefficients (b,u,v,w). To the latter we now attach an index $(\rho, \varepsilon, \alpha)$ in order to distinguish the type of tracking, and then obtain the following matrix equation:

$$
\begin{pmatrix} \delta\rho \\ \delta\varepsilon \\ \delta\alpha \end{pmatrix} = \begin{pmatrix} b_\rho & u_\rho & v_\rho & w_\rho \\ b_\varepsilon & u_\varepsilon & v_\varepsilon & w_\varepsilon \\ b_\alpha & u_\alpha & v_\alpha & w_\alpha \end{pmatrix} \begin{pmatrix} 1 \\ s - s_0 \\ \cos s \\ \sin s \end{pmatrix}
$$

The left hand sides of the two equations above are identical, so we equate both right hand sides and separate them according to their dependence on the variable s into the 4 parts: Constant, linear, cosine, sine. After some lengthy algebraic work we arrive at the result below. We will see later that the success of the analytical method is due to the fact that the equations split into two independent groups: One from the constant and linear terms and one from the cosine and sine terms. The same result was obtained in Section 8.6 by least squares fits.

The first part connects the coefficients from the constant and the linear terms $b + u\,(s - s_0)$ with the mean longitude at epoch λ_0 and the mean longitude drift rate D by means of a 2×6 matrix:

$$
\begin{pmatrix}
b_\rho \\
u_\rho \\
\rho b_\varepsilon \\
\rho u_\varepsilon \\
\rho \cos \varepsilon\, b_\alpha \\
\rho \cos \varepsilon\, u_\alpha
\end{pmatrix}
= A
\begin{pmatrix}
-\sin \beta \sin \gamma & -(2/3) \cos \beta \\
0 & -\sin \beta \sin \gamma \\
\cos \beta \sin \gamma & -(2/3) \sin \beta \\
0 & \cos \beta \sin \gamma \\
-\cos \gamma & 0 \\
0 & -\cos \gamma
\end{pmatrix}
\begin{pmatrix}
\delta\lambda_0 \\
\delta D
\end{pmatrix}
$$

The second part connects the coefficients from the cosine and sine terms $v \cos s + w \sin s$ with the eccentricity and inclination vectors $\bar{e}$ and $\bar{i}$. It contains a 4×6 matrix, but its appearance can be simplified to a 2×3 matrix by the use of complex numbers. We denote the imaginary unit by j ($j^2 = -1$), since i is used for the inclination.

$$
\begin{pmatrix}
v_\rho + j w_\rho \\
\rho\,(v_\varepsilon + j w_\varepsilon) \\
\rho \cos \varepsilon\,(v_\alpha + j w_\alpha)
\end{pmatrix}
=
$$

$$
= A
\begin{pmatrix}
-\cos \beta - 2j \sin \beta \sin \gamma & \sin \beta \cos \gamma \\
-\sin \beta + 2j \cos \beta \sin \gamma & -\cos \beta \cos \gamma \\
-2j \cos \gamma & -\sin \gamma
\end{pmatrix}
\begin{pmatrix}
\delta e_x + j \delta e_y \\
\delta i_x + j \delta i_y
\end{pmatrix}
$$

The importance of the equations derived above lies in the fact that they show how the 6 synchronous orbital elements (λ_0, D, e_x, e_y, i_x, i_y) can be derived in a simple way from the 4 coefficients (b, u, v, w) for different types of tracking. The determination errors of the coefficients were estimated in Section 8.6, and they can now be translated into errors in the determined orbital elements. The absence of observability of an element is indicated by an error that increases towards infinity.

I. Mean longitude and drift from one station ranging.
To start with we investigate the result of ranging from one ground station.
One can obtain the mean longitude and its drift rate as deterministic solutions, without any least squares fit, from the above equation. Below we
have defined a new dimension-less auxiliary parameter τ and moved the
unperturbed semimajor axis A to the left in order to obtain the along-track
position in km.

$$\tau = \frac{2 \cos \beta}{3 \sin \beta \sin \gamma}$$

$$A\delta\lambda_0 = \frac{\tau u_\rho - b_\rho}{\sin \beta \sin \gamma}$$

$$A\delta D = -\frac{u_\rho}{\sin \beta \sin \gamma}$$

The most important parameter above is the following dimension-less factor,
which can also be expressed according the definitions in Section 8.3

$$\frac{1}{\sin \beta \sin \gamma} = \frac{\rho}{R \sin(l - \lambda) \cos \phi}$$

It is plotted in Figure A as a function of ground station latitude (ϕ) and
longitude relative to the spacecraft ($l - \lambda$). With the help of Figure A or the
above equation, one can now estimate the expected error in the determination of the mean longitude and its drift rate from ranging.

We can insert the typical value $3\sigma_b = 15$ m, from Section 8.6. It shall be
multiplied by a factor, from Figure A, of at least 10, and in most cases below 100, although going to infinity to the left. This gives the mean along-
track error in $A\delta\lambda_0$ of between 150 m ($= 0.0002°$) and 1.5 km ($= 0.002°$)
with 99% probability. However, the error of the actual longitude is higher,
since it is obtained by adding the librations from the eccentricity as calculated in *V*.

When the station and spacecraft longitudes coincide, i.e. $l = \lambda$ the error becomes infinite, and it does not help that the latitudes are different. In that
situation, the longitude and its drift rate are not observable by ranging and
can only be determined by azimuth tracking, as described in *III*.

From the latest equations one can obtain the time evolution of the estimated error in the predicted mean longitude. At the time $t - t_0$, corresponding to the sidereal angle $s - s_0$, from the epoch (at s_0) the mean along-track position that is obtained from the mean longitude contains the following error with the covariance given below:

$$A\delta\lambda = A\delta\lambda_0 + A\delta D(s - s_0) = \frac{(\tau + s_0 - s)u_\rho - b_\rho}{\sin\beta\sin\gamma}$$

$$A^2\sigma_\lambda^2 = A^2cov(\delta\lambda, \delta\lambda) =$$

$$= \frac{(\tau + s_0 - s)^2cov(u_\rho, u_\rho) + 2(\tau + s_0 - s)cov(b_\rho, u_\rho) + cov(b_\rho, b_\rho)}{(\sin\beta\sin\gamma)^2}$$

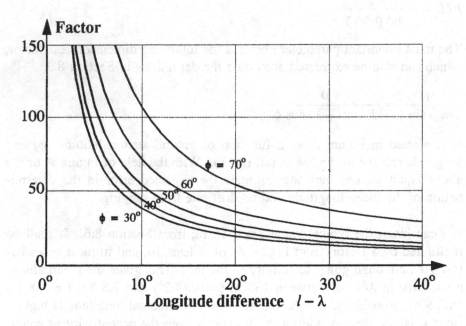

Figure 8.7.A. Dimension-less factor $1/(\sin\beta\sin\gamma)$ for converting ranging error to the minimum error in the determined along-track position that corresponds to the *mean* longitude. The function is anti-symmetric and has the same value, but negative, when $l - \lambda < 0$. The station latitude $\phi = 30°$, 40°, 50°, 60°, 70°.

It is always possible to select the epoch s_0 inside the tracking interval such that

$$cov(b_\rho, u_\rho) = 0$$

according to Section 8.6. The equivalent result from below can also be obtained with any other choice of epoch, but the derivation then becomes more cumbersome. We obtain the standard deviation of the error in the mean longitude at the sidereal angle $= s$

$$A\sigma_\lambda = \frac{\sqrt{(\tau + s_0 - s)^2 \sigma_u^2 + \sigma_b^2}}{|\sin \beta \sin \gamma|} \geq \frac{\sigma_b}{|\sin \beta \sin \gamma|}$$

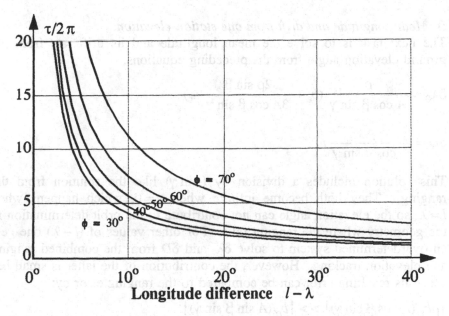

Figure 8.7.B. Time $= \tau/2\pi$ in sidereal days after epoch when the minimum error in the mean longitude occurs. The function is anti-symmetric and has the same value, but negative, when $l - \lambda < 0$. The station latitude $\phi = 30°$, $40°$, $50°$, $60°$, $70°$.

The surprising result here is that the longitude error is minimum, not at the epoch, but when $s = s_0 + \tau$. When the spacecraft longitude lies to the west of the station, τ is positive and the minimum may occur several days after the epoch and even after the end of the tracking interval. Figure B shows a plot of $\tau/2\pi$, which is the time of minimum error expressed in orbital revolutions = sidereal days, from the epoch.

This phenomenon of a decreasing error during part of the prediction time only happens when the spacecraft is located west of the station. When it lies to the east one gets a negative τ, so the time of minimum error is in the past. The anti-symmetry that makes the sign of τ the same as that of $(l - \lambda)$ is due to the direction of the Earth's rotation. In most cases the prediction accuracy is more important than the accuracy of the past, so there is an advantage in locating the spacecraft west of the station.

II. Mean longitude and drift from one station elevation.
The next task is to solve the mean longitude and its drift rate from the ground elevation angle from the preceding equations.

$$\delta\lambda_0 = \frac{\rho}{A \cos \beta \sin \gamma} \left(b_\varepsilon + \frac{2\rho \sin \beta}{3A \cos \beta \sin \gamma} u_\varepsilon\right)$$

$$\delta D = \frac{\rho}{A \cos \beta \sin \gamma} u_\varepsilon$$

This solution includes a division by $(\sin \gamma)$ like the solution from the ranging. They both become infinite when $\gamma = 0$, which happens when $l = \lambda$, so the elevation angle can not contribute to the orbit determination in the geometry where the ranging fails. For other values of $(l - \lambda)$ one gets an overdetermined system to solve $\delta\lambda_0$ and δD from the combined ranging and elevation tracking. However, the contribution of the latter is small because its resulting error can be compared to the ranging error by:

$$|\rho b_\varepsilon/(A \cos \beta \sin \gamma)| >> |b_\rho/(A \sin \beta \sin \gamma)|$$

In the inequality above, the left hand side is about 10 times greater than the right hand side. The contribution of the elevation to the determination of the mean longitude and its drift rate is meaningful mainly when there is no ranging data at all.

III. Mean longitude and drift from one station azimuth.

Of higher practical importance is the solution of the mean longitude and its drift rate from the ground azimuth tracking. We obtain from the preceding equations, after temporarily introducing the new parameter χ

$$\chi = \frac{\rho \cos \varepsilon}{A \cos \gamma} = \frac{1 - \cos^2(l - \lambda) \cos^2 \phi}{\sin \phi}$$

$$\delta \lambda_0 = - \chi b_\alpha \quad ; \quad \delta D = - \chi u_\alpha$$

This solution is singular only for an equatorial station ($\phi = 0$). It is very important that, for other values of ϕ, the solution does not become infinite when $l = \lambda$. The orbital elements λ_0 and D are then observable *only* by the azimuth tracking, whereas the other tracking types, range and elevation, cannot contribute anything to this determination. The azimuth measures directly the longitude, after multiplication by

$$- \chi = - \sin \phi \quad \text{(when } l = \lambda)$$

Any error in the azimuth, including a constant bias, is now directly translated into an error of the determined longitude, when $\lambda = l$. The only remedy against a bias is to check and calibrate regularly the azimuth measurement with ranging from a second station.

For other spacecraft longitudes ($\lambda \neq l$), the contribution of the azimuth for the determination of λ_0 and D is small in comparison with the ranging, like it was for the elevation. There is a smooth transition between the geometries with $\lambda = l$ and $\lambda \neq l$ that is shown in Figure C. It is similar to figure A, but the singularity has been removed by the inclusion of azimuth data by combining the two tracking equations by least squares with optimal weighting factors. The result is the error in $\delta \lambda_0$ at $s = s_0 + \tau$ with 99% probability:

- Range: $\delta \lambda_0 = - b_\rho /(A \sin \beta \sin \gamma)$ with $3\sigma_b = 15$ m

- Azimuth: $\delta \lambda_0 = - b_\alpha \chi$ with $3\sigma_b = 0.005°$

It is obvious from Figure C that, because of the determination accuracy, one should avoid placing the spacecraft and the ground station too close in longitude.

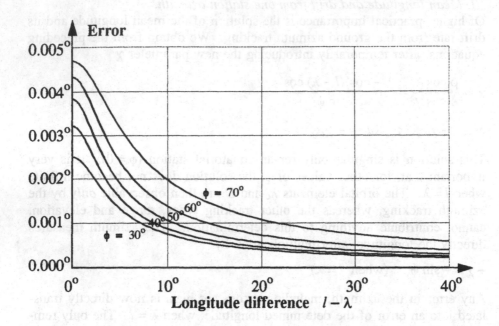

Figure 8.7.C. 99% error in the *mean* longitude assuming the 99% error (in *b*) of ranging = 15 m and of azimuth = 0.005°. The function is symmetric and has the same value when $l - \lambda < 0$. The station latitude ϕ = 30°, 40°, 50°, 60°, 70°.

IV. Eccentricity and inclination vectors from one station.

The next task is now to use the second matrix equation that was derived at the beginning of the section. The equation below connects the eccentricity and inclination vectors with the coefficients v for the cosine and w for the sine variations as functions of the spacecraft sidereal angle s for each type of tracking.

$$
\begin{pmatrix}
v_\rho + jw_\rho \\
\rho \, (v_\varepsilon + jw_\varepsilon) \\
\rho \cos \varepsilon \, (v_\alpha + jw_\alpha)
\end{pmatrix} =
$$

$$
= A \begin{pmatrix}
-\cos \beta - 2j \sin \beta \sin \gamma & \sin \beta \cos \gamma \\
-\sin \beta + 2j \cos \beta \sin \gamma & -\cos \beta \cos \gamma \\
-2j \cos \gamma & -\sin \gamma
\end{pmatrix}
\begin{pmatrix}
\delta e_x + j\delta e_y \\
\delta i_x + j\delta i_y
\end{pmatrix}
$$

If there is data from all three types of tracking, this becomes an overdetermined system for solving 2 complex variables $\delta\bar{e}$ and $\delta\bar{i}$ from 3 complex parameters. Or, equivalently, solve by least squares 4 normal variables from 6 input parameters.

$$(\delta e_x, \delta e_y, \delta i_x, \delta i_y) \quad \text{from} \quad (v_\rho, w_\rho, v_\varepsilon, w_\varepsilon, v_\alpha, w_\alpha)$$

One can see that any combination of 2 of the 3 tracking types $(\rho, \varepsilon, \alpha)$ provides a deterministic system of equations since the corresponding 2×2 complex sub-determinants of the matrix are not zero, not even when $\gamma = 0$. The only exception occurs when $\gamma = \pm 90°$, which corresponds to a ground station on the equator on a different longitude from the spacecraft. In this case, the two components of the inclination vector $\bar{i}$ can only be determined from the azimuth angle α.

A full orbit determination can be performed with the input consisting of the two antenna angles. On the other hand, only one type of tracking, e.g. ranging, from one ground station does not provide enough observability for both the eccentricity $(\delta\bar{e})$ and the inclination $(\delta\bar{i})$ vectors, although we have seen, in I, that it is sufficient for determining the mean longitude and its drift rate. Regardless of how long the tracking interval is and how much tracking is collected, one can still only determine 4 of the 6 orbital elements by one type of tracking of a geostationary spacecraft.

Instead of solving the overdetermined 2×3 system of equations above by least squares, we now split it into two deterministic 2×2 systems in order to better illustrate the obtained orbit accuracy. In the first system we combine range and elevation and in the second range and azimuth.

Ranging and elevation give:

$$\begin{pmatrix} v_\rho + jw_\rho \\ \rho\,(v_\varepsilon + jw_\varepsilon) \end{pmatrix} =$$

$$= A \begin{pmatrix} -\cos\beta - 2j\sin\beta\sin\gamma & \sin\beta\cos\gamma \\ -\sin\beta + 2j\cos\beta\sin\gamma & -\cos\beta\cos\gamma \end{pmatrix} \begin{pmatrix} \delta e_x + j\delta e_y \\ \delta i_x + j\delta i_y \end{pmatrix}$$

Ranging and azimuth give:

$$\begin{pmatrix} v_\rho + jw_\rho \\ \rho\cos\varepsilon\,(v_\alpha + jw_\alpha) \end{pmatrix} =$$

$$= A \begin{pmatrix} -\cos\beta - 2j\sin\beta\sin\gamma & \sin\beta\cos\gamma \\ -2j\cos\gamma & -\sin\gamma \end{pmatrix} \begin{pmatrix} \delta e_x + j\delta e_y \\ \delta i_x + j\delta i_y \end{pmatrix}$$

These two 2×2 linear complex sets of equations can easily be solved analytically. We can simplify them even further by approximating on the right hand side

$$v_\rho + jw_\rho \approx 0$$

since the ranging error is vanishingly small in comparison with the angle errors. This is the case for all longitudes, because here the contribution from the ranging does not become singular when $\lambda = l$.

In the solution we also take the absolute value of the complex expressions on both sides due to the fact that the individual components of the errors are not of interest but only the magnitudes:

$$|\delta\bar{e}| = \sqrt{\delta e_x^2 + \delta e_y^2} \qquad |\delta\bar{i}| = \sqrt{\delta i_x^2 + \delta i_y^2}$$

The determination error from the elevation, when combined with ranging, now becomes:

$$|\delta\bar{e}| = \frac{\rho\sin\beta}{A}\sqrt{v_\varepsilon^2 + w_\varepsilon^2}$$

$$|\delta\bar{i}| = \frac{\rho\sqrt{\cos^2\beta + 4\sin^2\beta\sin^2\gamma}}{A|\cos\gamma|}\sqrt{v_\varepsilon^2 + w_\varepsilon^2}$$

The determination error from the azimuth, when combined with ranging, becomes:

$$|\delta\bar{e}| = \frac{\rho\cos\varepsilon\sin\beta|\cos\gamma|}{A\sqrt{\cos^2\beta\sin^2\gamma + 4\sin^2\beta}}\sqrt{v_\alpha^2 + w_\alpha^2}$$

$$|\delta\bar{i}| = \frac{\rho\cos\varepsilon\sqrt{\cos^2\beta + 4\sin^2\beta\sin^2\gamma}}{A\sqrt{\cos^2\beta\sin^2\gamma + 4\sin^2\beta}}\sqrt{v_\alpha^2 + w_\alpha^2}$$

The dimension-less factors in front of the error amplitude $\sqrt{v^2 + w^2}$ in the 4 equations above are plotted, for the eccentricity in Figure D and the inclination in Figure E. With these factors one shall multiply the error amplitudes σ_{vw} for azimuth and elevation from one of the four figures in

Section 8.6 in order to obtain an estimate of the accuracy of a determination with the corresponding tracking configuration. The result gives the contribution of the azimuth and elevation errors separately, when combined with ranging. All three trackings combined in an optimally weighted least squares fit would produce a result that were slightly better than the best of the two separate results, from (ε, ρ) and (α, ρ). In almost all situations, the factor for the elevation is smaller than for the azimuth, which compensates for the higher elevation error.

It is interesting to note that only the librations (with a period of one sidereal day) of the tracking data is used for the determination of the 4 elements $(\delta e_x, \delta e_y, \delta i_x, \delta i_y)$ or, equivalently, (e, i, Ω, ω). A tracking error with another time dependence or a constant bias has no influence on the accuracy of these elements when the tracking interval is long enough and a sufficient volume of tracking data is available.

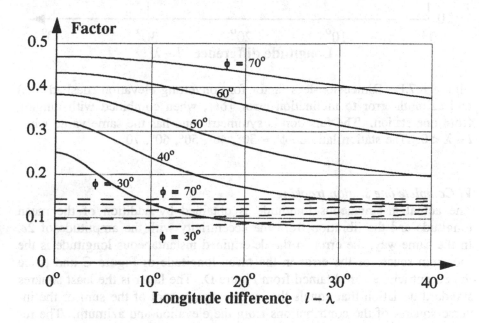

Figure 8.7.D. Dimension-less factor for converting elevation (dashed line) and azimuth error to eccentricity error $|\delta \bar{e}|$, when combined with ranging from one station. The function is symmetric and has the same value when $l - \lambda < 0$. The station latitude $\phi = 30°, 40°, 50°, 60°, 70°$.

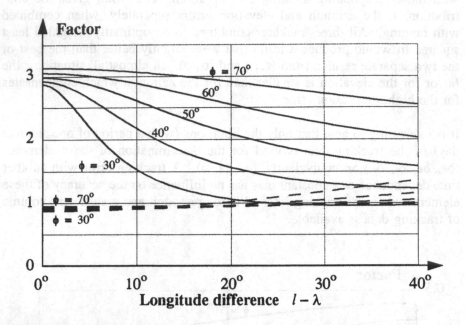

Figure 8.7.E. Dimension-less factor for converting elevation (dashed line) and azimuth error to inclination error $|\delta \bar{i}|$, when combined with ranging from one station. The function is symmetric and has the same value when $l - \lambda < 0$. The station latitude ϕ = 30°, 40°, 50°, 60°, 70°.

V. Complete one station tracking.
The actual instantaneous longitude is obtained by addition of the mean longitude and the libration from the eccentricity with the amplitude of $2e$. In the same way, the error in the determined instantaneous longitude is the root sum square of the error in the mean longitude of Figure C and twice the eccentricity error obtained from Figure D. The latter is the least squares standard deviation that equals the inverse square root of the sum of the inverse squares of the contributions from the elevation and azimuth. The resulting 99% error is shown in Figure F with the assumed 99% angle errors:

- Elevation: $3\sigma_{vw} = 0.006°$

- Azimuth: $3\sigma_{vw} = 0.004°$

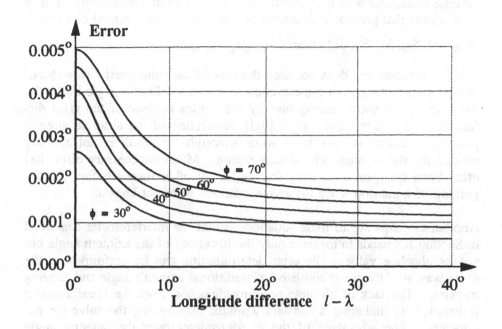

Figure 8.7.F. 99% error in the actual longitude assuming the errors (in b) of ranging and azimuth as in Figure C. The error from the librations caused by the eccentricity is added, determined from 99% amplitude errors (in v,w) for elevation = 0.006° and for azimuth = 0.004°. The function is symmetric and has the same value when $l - \lambda < 0$. The station latitude ϕ = 30°, 40°, 50°, 60°, 70°.

An important consequence of the combined results of I and IV is the fact that with tracking from range, azimuth and elevation from one station there is a certain observability redundancy. When there is sufficient longitude separation, both λ_0 and D are observable by any one of the tracking types ranging, azimuth or elevation, but since the ranging is so much more accurate it will dominate the solution. When combined with ranging, the orbit can be determined from partial information of one angle. Only the two coefficients (v,w) or, equivalently, amplitude and phase, of the libration with one sidereal day period are needed from one of the antenna angles.

A very useful tracking configuration consists of all the 4 coefficients from ranging combined with the 2 coefficients (v_α, w_α) from the azimuth; in all 6 coefficients that provide a deterministic solution of the 6 orbital elements:

$$(\delta\lambda_0, \delta D, \delta e_x, \delta e_y, \delta i_x, \delta i_y) \quad \text{from} \quad (b_\rho, u_\rho, v_\rho, w_\rho, v_\alpha, w_\alpha)$$

For this solution one does not need the other 2 azimuth coefficients (b_α, u_α) or any contribution from the elevation. One can calibrate angle biases by the ranging but not a ranging bias by the angles because of the great difference in accuracy. Such an in-flight calibration of the absolute angular pointing direction is easy to perform although the result is not actually needed for the nominal orbit determination. Much unnecessary effort has often been spent on calibrating the angles by other means before the beginning of a mission when the above relations were not known.

Another consequence of these equations is that the interferometer described in Section 8.2 needs to measure only the librations of the azimuth angle but not its absolute value. The orbit determination can be performed in the same way as if the input consisted of traditional azimuth angle and ranging tracking. The lack of absolute pointing direction from the interferometer is handled by including a constant azimuth bias among the solve-for parameters. The advantage of the interferometer over the antenna angle measurements is expected to be an improvement in the accuracy by at least an order of magnitude.

If the antenna is built on soft ground or in an area with geodetic activity it might cause the angle biases to be slowly changing. This "tower-of-Pisa-effect" was experienced with an antenna at Fucino (Italy) that tracked ESA's Olympus spacecraft between 1989 and 1993, but the orbit accuracy was not adversely affected during the first half of the mission. With the wide longitude separation $l - \lambda = 32.6°$, ranging provided most of the tracking input and only the librations of the angles were needed for the orbit determination. However, during the second half of the mission the orbit was strongly perturbed by attitude control thrusts needed for compensating some on-board failures. These unmodelled perturbations made it difficult to fit an orbit through the tracking measurements, which to a great extent invalidated the present analytical theory. The orbit determination had to be replaced, to a large extent, by instantaneous position determination through range, azimuth and elevation as shown in Figure 8.3.A.

VI. Two station ranging.

The next task is to examine the tracking configuration with ranging from two ground stations. If there is, in addition, angle tracking from one or both of these stations, its contribution to the orbit determination accuracy is negligible, compared to the ranging. Determination of the two elements (λ_0, D) can be done with the ranging from any one of the two stations according to the method of *I*, provided that it does not lie on the same longitude as the spacecraft.

If none of the stations lie on the spacecraft longitude there is a tracking redundancy for the determination of (λ_0, D). One can perform the determination by a least squares fit or calibrate a bias of one ranging system by means of the other one. In the early operations with ESA's test spacecraft OTS-2 in 1978 there was a constant discrepancy of 1.5 km between ranging measurements from Fucino and from Villafranca (Spain). In order to settle the question which station was right, it was necessary to temporarily range from a third station, in England. By the majority principle it was then decided that Fucino was right.

The equation to determine the 4 remaining elements (e_x, e_y, i_x, i_y) from ranging by two stations can be obtained from the preceding single station matrix equation. We extract the row with the ranging coefficients $(v_\rho + jw_\rho)$ and repeat it for the second station. Since all measurements are made in range, we can omit the index ρ and instead introduce the index $k = 1$ and 2 for the two stations of the coefficients (v, w) and of β and γ. For simplicity we introduce the following two new abbreviations:

$$J_k = -\cos \beta_k - 2j \sin \beta_k \sin \gamma_k$$

$$E_k = \sin \beta_k \cos \gamma_k$$

The 2×2 complex system of equations to solve the eccentricity and inclination vectors from two station ranging now becomes:

$$v_k + jw_k = A[J_k (\delta e_x + j\delta e_y) + E_k (\delta i_x + j\delta i_y)] \quad \text{with } k = 1,2$$

We denote here the determinant of the system by the new parameter

$$H = J_1 E_2 - J_2 E_1 =$$

$$= \sin \beta_1 \cos \beta_2 \cos \gamma_1 - \sin \beta_2 \cos \beta_1 \cos \gamma_2 + 2j \sin \beta_1 \sin \beta_2 \sin(\gamma_2 - \gamma_1)$$

The determinant is zero only when either both stations coincide ($\beta_1 = \beta_2$ and $\gamma_1 = \gamma_2$) or both lie anywhere on the equator ($\gamma_1 = \pm 90°$ and $\gamma_2 = \pm 90°$). In the latter case, the eccentricity is still observable because also both $E_k = 0$, but this is not of any practical significance as long a the inclination error is infinite.

From the system of equations we now obtain the following deterministic solution for $\delta \bar{e}$ and $\delta \bar{i}$ when $H \neq 0$.

$$\delta e_x + j\delta e_y = [E_2 (v_1 + jw_1) - E_1 (v_2 + jw_2)]/AH$$

$$\delta i_x + j\delta i_y = [-J_2 (v_1 + jw_1) + J_1 (v_2 + jw_2)]/AH$$

An upper bound of the determination errors can be found by:

$$|\delta \bar{e}| \leq \frac{|E_1| + |E_2|}{A|H|} \max_{k=1,2} \sqrt{v_k^2 + w_k^2}$$

$$|\delta \bar{i}| \leq \frac{|J_1| + |J_2|}{A|H|} \max_{k=1,2} \sqrt{v_k^2 + w_k^2}$$

However, a better measure of the determination error is the standard deviation below. We assume that the ranging errors from the two stations are of the same size, but uncorrelated, with the following standard deviation, for $k = 1$ and 2:

$$\sigma_{vw} = 5 \text{ m} \quad \text{for} \quad \sqrt{v_k^2 + w_k^2}$$

The standard deviations of the eccentricity and inclination errors now become:

$$\sigma_e = \frac{\sqrt{E_1^2 + E_2^2}}{|H|} \frac{\sigma_{vw}}{A}$$

$$\sigma_i = \frac{\sqrt{|J_1|^2 + |J_2|^2}}{|H|} \frac{\sigma_{vw}}{A}$$

The dimension-less factors in front of σ_{vw}/A can easily be calculated by pure geometry when the longitudes and latitudes of the two stations are known, but they are difficult to visualise graphically as functions of 4 independent variables. The best accuracy is obtained by one station in the northern hemisphere and the other one in the southern, as far apart in latitude and longitude as the visibility allows. With a minimum elevation of 20° one obtains the optimal factors 0.7 and 5 for the eccentricity and inclination, respectively.

$$\sigma_e \geq 0.7\,\sigma_{vw}/A \quad ; \quad \sigma_i \geq 5\,\sigma_{vw}/A$$

As an example of a less favourable but more typical geometry, we calculate the factors 13 and 100 when both stations lie on latitude 50° north but with the longitudes 15° on either side of the spacecraft. The 99% error level equals now, when

$$\phi_1 = \phi_2 = 50° \quad ; \quad l_1 = \lambda - 15° \quad ; \quad l_2 = \lambda + 15°$$

$$3\sigma_e = 13 \times 3\sigma_{vw}/A = 4.6 \times 10^{-6}$$

$$3\sigma_i = 100 \times 3\sigma_{vw}/A = 3.6 \times 10^{-5}\,\text{rad} = 0.002°$$

VII. Three station ranging.
It is relatively unusual to use three ranging stations in a geostationary mission. With three simultaneous measurements one can determine the instantaneous spacecraft position by pure geometrical *trilateration*, without referring to the orbital propagation model. Still, a propagated orbit ahead of time is needed for station keeping and, for most missions, also by the payload operations and for ground antenna predictions. The orbital elements and other solve-for parameters are then calculated by fitting the propagated orbit through the determined positions.

The position in the Earth-rotating co-ordinate system of the spacecraft is here denoted by $\bar{r}$ and of the three ground stations by $\bar{R}_k$ with $k = 1,2,3$. They are calculated from the equations in Section 8.1. The three vectors from the stations to the spacecraft

$$\bar{\rho}_k = \bar{r} - \bar{R}_k \quad \text{with } k = 1,2,3$$

are, in magnitude, equal to the three ranging measurements:

$$\rho_k = |\bar{\rho}_k| = |\bar{r} - \bar{R}_k| \quad ; \quad k = 1,2,3$$

After squaring the above equation

$$\rho_k^2 = \bar{r} \cdot \bar{r} - 2\bar{r} \cdot \bar{R}_k + R_k^2 \quad ; \quad k = 1,2,3$$

one can take the differential of the spacecraft position with respect to the ranging measurements:

$$2\rho_k\, \delta\rho_k = 2\bar{r}\cdot\delta\bar{r} - 2\delta\bar{r}\cdot\bar{R}_k \;\; ; \;\;\; k = 1,2,3$$

$$\rho_k\, \delta\rho_k = \bar{\rho}_k\cdot\delta\bar{r} \;\; ; \;\;\; k = 1,2,3$$

The last equation above is a 3×3 linear system, which after matrix inversion provides a solution in the form below, where the three vectors $\bar{c}_j$ are the columns of the inverse of the matrix with rows $=\bar{\rho}_k/\rho$

$$\delta\bar{r} = \bar{c}_1\delta\rho_1 + \bar{c}_2\delta\rho_2 + \bar{c}_3\delta\rho_3$$

The typical ranging error $\delta\rho$ here is not necessarily the same as the previous errors, which were estimated from the residuals after fitting measurements to the orbital model. However, as an order of magnitude we can still assume the standard deviation of all the three measurement errors to be the same, namely $\sigma_\rho = 5$ m, but uncorrelated. The standard deviation of the position error is then:

$$\sigma_r = \sigma_\rho\sqrt{c_1^2 + c_2^2 + c_3^2}$$

Figure G shows an example of a favourable and of an unfavourable geometry for ranging a spacecraft at longitude $\lambda = 0°$ from three stations. The 99% error levels are, respectively:

$3\sigma_r = 7.4 \times 3\sigma_\rho = 0.11$ km from:

Stockholm:	$l = +18.1°$	$\phi = +59.2°$	$\varepsilon = 21°$
Kourou:	$l = -52.8°$	$\phi = +5.2°$	$\varepsilon = 29°$
Mauritius:	$l = +57.5°$	$\phi = -20.3°$	$\varepsilon = 22°$

$3\sigma_r = 84 \times 3\sigma_\rho = 1.2$ km from:

Darmstadt:	$l = +9.0°$	$\phi = +49.7°$	$\varepsilon = 32°$
Villafranca:	$l = -3.9°$	$\phi = +40.4°$	$\varepsilon = 43°$
Fucino:	$l = +13.6°$	$\phi = +42.0°$	$\varepsilon = 40°$

This instantaneous determination accuracy can be improved by fitting a sequence of calculated orbit positions to the orbit model. The result will then be about the same as a combination of three one-station results from *I* with two two-station results from *VI* by least squares.

Figure 8.7.G. The wide triangle Stockholm-Kourou-Mauritius provides 11 times better accuracy in trilateration than the small, shaded, triangle Darmstadt-Villafranca-Fucino for a spacecraft at $\lambda = 0°$.

TABLES

year	G	year	G	days/year
1980	99.814	2012	100.060	366
1981	100.561	2013	100.807	365
1982	100.322	2014	100.568	365
1983	100.083	2015	100.330	365
1984	99.845	2016	100.091	366
1985	100.592	2017	100.838	365
1986	100.353	2018	100.599	365
1987	100.114	2019	100.361	365
1988	99.875	2020	100.122	366
1989	100.622	2021	100.869	365
1990	100.384	2022	100.630	365
1991	100.145	2023	100.391	365
1992	99.906	2024	100.153	366
1993	100.653	2025	100.900	365
1994	100.414	2026	100.661	365
1995	100.176	2027	100.422	365
1996	99.937	2028	100.183	366
1997	100.684	2029	100.930	365
1998	100.445	2030	100.692	365
1999	100.207	2031	100.453	365
2000	99.968	2032	100.214	366
2001	100.715	2033	100.961	365
2002	100.476	2034	100.722	365
2003	100.237	2035	100.484	365
2004	99.999	2036	100.245	366
2005	100.746	2037	100.992	365
2006	100.507	2038	100.753	365
2007	100.268	2039	100.515	365
2008	100.029	2040	100.276	366
2009	100.776	2041	101.023	365
2010	100.538	2042	100.784	365
2011	100.299	2043	100.545	365

Table 1. Sidereal angle G (degrees) of the Greenwich meridian at 0 hour UTC on each January 1. The number of days in the year is also listed, with every 4 years a leap-year, including the year 2000.

Table 2   287

λ	B	$\ddot{\lambda}$	ΔV	λ	B	$\ddot{\lambda}$	ΔV
0	-21.3	0.65	0.67	+1	-23.1	0.70	0.73
-1	-19.5	0.59	0.61	+2	-24.9	0.76	0.79
-2	-17.6	0.54	0.56	+3	-26.7	0.81	0.84
-3	-15.8	0.48	0.50	+4	-28.4	0.87	0.90
-4	-13.9	0.42	0.44	+5	-30.2	0.92	0.95
-5	-12.1	0.37	0.38	+6	-31.9	0.97	1.01
-6	-10.2	0.31	0.32	+7	-33.5	1.02	1.06
-7	-8.4	0.25	0.26	+8	-35.2	1.07	1.11
-8	-6.5	0.20	0.21	+9	-36.8	1.12	1.16
-9	-4.6	0.14	0.15	+10	-38.3	1.17	1.21
-10	-2.8	0.09	0.09	+11	-39.8	1.21	1.26
-11	-1.0	0.03	0.03	+12	-41.3	1.26	1.30
-12	0.9	-0.03	0.03	+13	-42.8	1.30	1.35
-13	2.7	-0.08	0.08	+14	-44.2	1.34	1.39
-14	4.5	-0.14	0.14	+15	-45.5	1.38	1.44
-15	6.3	-0.19	0.20	+16	-46.8	1.42	1.48
-16	8.0	-0.24	0.25	+17	-48.0	1.46	1.52
-17	9.8	-0.30	0.31	+18	-49.2	1.50	1.55
-18	11.5	-0.35	0.36	+19	-50.3	1.53	1.59
-19	13.2	-0.40	0.42	+20	-51.4	1.56	1.62
-20	14.8	-0.45	0.47	+21	-52.4	1.59	1.65
-21	16.5	-0.50	0.52	+22	-53.3	1.62	1.68
-22	18.1	-0.55	0.57	+23	-54.2	1.65	1.71
-23	19.6	-0.60	0.62	+24	-55.0	1.67	1.74
-24	21.2	-0.64	0.67	+25	-55.7	1.70	1.76
-25	22.7	-0.69	0.72	+26	-56.4	1.72	1.78
-26	24.2	-0.74	0.76	+27	-57.0	1.73	1.80
-27	25.6	-0.78	0.81	+28	-57.5	1.75	1.81
-28	27.0	-0.82	0.85	+29	-58.0	1.76	1.83
-29	28.3	-0.86	0.89	+30	-58.3	1.77	1.84

Table 2. As a function of the longitude (λ, in degrees East) is listed the acceleration (B, in 10^{-9} m/s^2), longitude change ($\ddot{\lambda}$ in 0.001°/day^2) and ΔV (in m/s per year) for longitude station keeping from the Earth's gravity field.

Table 2

λ	B	λ̈	ΔV	λ	B	λ̈	ΔV
-30	29.6	-0.90	0.93	+31	-58.6	1.78	1.85
-31	30.9	-0.94	0.97	+32	-58.8	1.79	1.86
-32	32.1	-0.98	1.01	+33	-59.0	1.79	1.86
-33	33.3	-1.01	1.05	+34	-59.0	1.80	1.86
-34	34.4	-1.05	1.09	+35	-59.0	1.80	1.86
-35	35.5	-1.08	1.12	+36	-58.9	1.79	1.86
-36	36.5	-1.11	1.15	+37	-58.7	1.79	1.85
-37	37.5	-1.14	1.18	+38	-58.5	1.78	1.84
-38	38.5	-1.17	1.21	+39	-58.1	1.77	1.83
-39	39.4	-1.20	1.24	+40	-57.7	1.76	1.82
-40	40.2	-1.22	1.27	+41	-57.2	1.74	1.80
-41	41.0	-1.25	1.29	+42	-56.6	1.72	1.79
-42	41.7	-1.27	1.32	+43	-55.9	1.70	1.76
-43	42.4	-1.29	1.34	+44	-55.2	1.68	1.74
-44	43.1	-1.31	1.36	+45	-54.3	1.65	1.71
-45	43.7	-1.33	1.38	+46	-53.4	1.63	1.69
-46	44.2	-1.35	1.40	+47	-52.4	1.60	1.65
-47	44.7	-1.36	1.41	+48	-51.4	1.56	1.62
-48	45.1	-1.37	1.42	+49	-50.2	1.53	1.59
-49	45.5	-1.39	1.44	+50	-49.0	1.49	1.55
-50	45.9	-1.40	1.45	+51	-47.7	1.45	1.51
-51	46.2	-1.40	1.46	+52	-46.4	1.41	1.46
-52	46.4	-1.41	1.46	+53	-44.9	1.37	1.42
-53	46.6	-1.42	1.47	+54	-43.4	1.32	1.37
-54	46.7	-1.42	1.47	+55	-41.8	1.27	1.32
-55	46.8	-1.42	1.48	+56	-40.2	1.22	1.27
-56	46.8	-1.43	1.48	+57	-38.5	1.17	1.22
-57	46.8	-1.43	1.48	+58	-36.7	1.12	1.16
-58	46.8	-1.42	1.48	+59	-34.9	1.06	1.10
-59	46.7	-1.42	1.47	+60	-33.0	1.01	1.04

Table 2, continued. As a function of the longitude (λ, in degrees East) is listed the acceleration (B, in 10^{-9} m/s²), longitude change (λ̈ in $0.001°$/day²) and ΔV (in m/s per year) for longitude station keeping from the Earth's gravity field.

Table 2   289

λ	B	λ̈	ΔV	λ	B	λ̈	ΔV
-60	46.5	-1.42	1.47	+61	-31.1	0.95	0.98
-61	46.3	-1.41	1.46	+62	-29.1	0.89	0.92
-62	46.1	-1.40	1.45	+63	-27.1	0.82	0.85
-63	45.8	-1.39	1.44	+64	-25.0	0.76	0.79
-64	45.4	-1.38	1.43	+65	-22.9	0.70	0.72
-65	45.0	-1.37	1.42	+66	-20.7	0.63	0.65
-66	44.6	-1.36	1.41	+67	-18.5	0.56	0.58
-67	44.1	-1.34	1.39	+68	-16.3	0.50	0.51
-68	43.6	-1.33	1.38	+69	-14.0	0.43	0.44
-69	43.0	-1.31	1.36	+70	-11.7	0.36	0.37
-70	42.4	-1.29	1.34	+71	-9.4	0.29	0.30
-71	41.8	-1.27	1.32	+72	-7.1	0.22	0.22
-72	41.1	-1.25	1.30	+73	-4.7	0.14	0.15
-73	40.3	-1.23	1.27	+74	-2.4	0.07	0.07
-74	39.6	-1.20	1.25	+75	0.0	0.00	0.00
-75	38.7	-1.18	1.22	+76	2.4	-0.07	0.08
-76	37.9	-1.15	1.20	+77	4.8	-0.15	0.15
-77	37.0	-1.13	1.17	+78	7.2	-0.22	0.23
-78	36.1	-1.10	1.14	+79	9.6	-0.29	0.30
-79	35.1	-1.07	1.11	+80	11.9	-0.36	0.38
-80	34.1	-1.04	1.08	+81	14.3	-0.44	0.45
-81	33.1	-1.01	1.04	+82	16.7	-0.51	0.53
-82	32.0	-0.97	1.01	+83	19.0	-0.58	0.60
-83	30.9	-0.94	0.98	+84	21.3	-0.65	0.67
-84	29.8	-0.91	0.94	+85	23.6	-0.72	0.74
-85	28.6	-0.87	0.90	+86	25.8	-0.79	0.82
-86	27.4	-0.83	0.87	+87	28.1	-0.85	0.89
-87	26.2	-0.80	0.83	+88	30.2	-0.92	0.95
-88	24.9	-0.76	0.79	+89	32.4	-0.99	1.02
-89	23.7	-0.72	0.75	+90	34.5	-1.05	1.09

Table 2, continued. As a function of the longitude (λ, in degrees East) is listed the acceleration (B, in 10^{-9} m/s²), longitude change (λ̈ in 0.001°/day²) and ΔV (in m/s per year) for longitude station keeping from the Earth's gravity field.

Table 2

λ	B	λ̈	ΔV	λ	B	λ̈	ΔV
-90	22.3	-0.68	0.71	+91	36.5	-1.11	1.15
-91	21.0	-0.64	0.66	+92	38.6	-1.17	1.22
-92	19.7	-0.60	0.62	+93	40.5	-1.23	1.28
-93	18.3	-0.56	0.58	+94	42.4	-1.29	1.34
-94	16.9	-0.51	0.53	+95	44.2	-1.35	1.40
-95	15.4	-0.47	0.49	+96	46.0	-1.40	1.45
-96	14.0	-0.43	0.44	+97	47.7	-1.45	1.51
-97	12.5	-0.38	0.40	+98	49.4	-1.50	1.56
-98	11.0	-0.34	0.35	+99	51.0	-1.55	1.61
-99	9.5	-0.29	0.30	+100	52.5	-1.60	1.66
-100	8.0	-0.24	0.25	+101	53.9	-1.64	1.70
-101	6.5	-0.20	0.21	+102	55.3	-1.68	1.74
-102	5.0	-0.15	0.16	+103	56.6	-1.72	1.78
-103	3.4	-0.10	0.11	+104	57.8	-1.76	1.82
-104	1.8	-0.06	0.06	+105	58.9	-1.79	1.86
-105	0.3	-0.01	0.01	+106	59.9	-1.82	1.89
-106	-1.3	0.04	0.04	+107	60.9	-1.85	1.92
-107	-2.9	0.09	0.09	+108	61.7	-1.88	1.95
-108	-4.5	0.14	0.14	+109	62.5	-1.90	1.97
-109	-6.1	0.18	0.19	+110	63.2	-1.92	2.00
-110	-7.7	0.23	0.24	+111	63.8	-1.94	2.01
-111	-9.2	0.28	0.29	+112	64.3	-1.96	2.03
-112	-10.8	0.33	0.34	+113	64.8	-1.97	2.04
-113	-12.4	0.38	0.39	+114	65.1	-1.98	2.05
-114	-14.0	0.43	0.44	+115	65.4	-1.99	2.06
-115	-15.6	0.47	0.49	+116	65.5	-1.99	2.07
-116	-17.1	0.52	0.54	+117	65.6	-2.00	2.07
-117	-18.7	0.57	0.59	+118	65.6	-2.00	2.07
-118	-20.2	0.62	0.64	+119	65.5	-1.99	2.07
-119	-21.8	0.66	0.69	+120	65.3	-1.99	2.06

Table 2, continued. As a function of the longitude (λ, in degrees East) is listed the acceleration (B, in 10^{-9} m/s^2), longitude change ($\ddot{\lambda}$ in $0.001°$/day^2) and ΔV (in m/s per year) for longitude station keeping from the Earth's gravity field.

Table 2   291

λ	B	$\ddot{\lambda}$	ΔV	λ	B	$\ddot{\lambda}$	ΔV
-120	-23.3	0.71	0.73	+121	65.0	-1.98	2.05
-121	-24.8	0.75	0.78	+122	64.6	-1.97	2.04
-122	-26.2	0.80	0.83	+123	64.1	-1.95	2.02
-123	-27.7	0.84	0.87	+124	63.6	-1.93	2.01
-124	-29.1	0.89	0.92	+125	62.9	-1.92	1.99
-125	-30.5	0.93	0.96	+126	62.2	-1.89	1.96
-126	-31.9	0.97	1.01	+127	61.4	-1.87	1.94
-127	-33.3	1.01	1.05	+128	60.6	-1.84	1.91
-128	-34.6	1.05	1.09	+129	59.6	-1.81	1.88
-129	-35.9	1.09	1.13	+130	58.6	-1.78	1.85
-130	-37.2	1.13	1.17	+131	57.5	-1.75	1.81
-131	-38.4	1.17	1.21	+132	56.3	-1.71	1.78
-132	-39.6	1.21	1.25	+133	55.0	-1.67	1.74
-133	-40.8	1.24	1.29	+134	53.7	-1.63	1.69
-134	-41.9	1.27	1.32	+135	52.3	-1.59	1.65
-135	-43.0	1.31	1.36	+136	50.9	-1.55	1.61
-136	-44.0	1.34	1.39	+137	49.4	-1.50	1.56
-137	-45.0	1.37	1.42	+138	47.8	-1.45	1.51
-138	-46.0	1.40	1.45	+139	46.2	-1.40	1.46
-139	-46.9	1.43	1.48	+140	44.5	-1.35	1.40
-140	-47.7	1.45	1.51	+141	42.8	-1.30	1.35
-141	-48.5	1.48	1.53	+142	41.0	-1.25	1.29
-142	-49.3	1.50	1.56	+143	39.1	-1.19	1.24
-143	-50.0	1.52	1.58	+144	37.3	-1.13	1.18
-144	-50.7	1.54	1.60	+145	35.4	-1.08	1.12
-145	-51.3	1.56	1.62	+146	33.4	-1.02	1.05
-146	-51.8	1.58	1.63	+147	31.5	-0.96	0.99
-147	-52.3	1.59	1.65	+148	29.4	-0.90	0.93
-148	-52.7	1.60	1.66	+149	27.4	-0.83	0.86
-149	-53.1	1.62	1.68	+150	25.3	-0.77	0.80

Table 2, continued. As a function of the longitude (λ, in degrees East) is listed the acceleration (B, in 10^{-9} m/s^2), longitude change ($\ddot{\lambda}$ in 0.001°/day^2) and ΔV (in m/s per year) for longitude station keeping from the Earth's gravity field.

Table 2

λ	B	λ̈	ΔV	λ	B	λ̈	ΔV
-150	-53.4	1.63	1.69	+151	23.3	-0.71	0.73
-151	-53.7	1.63	1.69	+152	21.2	-0.64	0.67
-152	-53.9	1.64	1.70	+153	19.0	-0.58	0.60
-153	-54.0	1.64	1.70	+154	16.9	-0.51	0.53
-154	-54.1	1.65	1.71	+155	14.8	-0.45	0.47
-155	-54.1	1.65	1.71	+156	12.6	-0.38	0.40
-156	-54.1	1.65	1.71	+157	10.5	-0.32	0.33
-157	-54.0	1.64	1.70	+158	8.3	-0.25	0.26
-158	-53.8	1.64	1.70	+159	6.2	-0.19	0.19
-159	-53.6	1.63	1.69	+160	4.0	-0.12	0.13
-160	-53.3	1.62	1.68	+161	1.9	-0.06	0.06
-161	-52.9	1.61	1.67	+162	-0.3	0.01	0.01
-162	-52.5	1.60	1.66	+163	-2.4	0.07	0.08
-163	-52.0	1.58	1.64	+164	-4.5	0.14	0.14
-164	-51.4	1.56	1.62	+165	-6.6	0.20	0.21
-165	-50.8	1.55	1.60	+166	-8.7	0.26	0.27
-166	-50.1	1.53	1.58	+167	-10.7	0.33	0.34
-167	-49.4	1.50	1.56	+168	-12.7	0.39	0.40
-168	-48.6	1.48	1.53	+169	-14.7	0.45	0.46
-169	-47.7	1.45	1.50	+170	-16.7	0.51	0.53
-170	-46.8	1.42	1.48	+171	-18.6	0.57	0.59
-171	-45.8	1.39	1.44	+172	-20.5	0.62	0.65
-172	-44.7	1.36	1.41	+173	-22.4	0.68	0.71
-173	-43.6	1.33	1.38	+174	-24.2	0.74	0.76
-174	-42.4	1.29	1.34	+175	-26.0	0.79	0.82
-175	-41.2	1.25	1.30	+176	-27.7	0.84	0.87
-176	-39.9	1.21	1.26	+177	-29.4	0.89	0.93
-177	-38.6	1.17	1.22	+178	-31.0	0.94	0.98
-178	-37.2	1.13	1.17	+179	-32.7	0.99	1.03
-179	-35.7	1.09	1.13	+180	-34.2	1.04	1.08

Table 2, continued. As a function of the longitude (λ, in degrees East) is listed the acceleration (B, in 10^{-9} m/s^2), longitude change ($\ddot{\lambda}$ in $0.001°/$day^2) and ΔV (in m/s per year) for longitude station keeping from the Earth's gravity field.

Table 3   293

Year	i_x	i_y	i	ΔV	Year	i_x	i_y	i	ΔV
1980	0.782	0.043	0.783	42.0	2003	0.916	0.063	0.918	49.1
1981	0.813	0.068	0.816	43.6	2004	0.942	0.026	0.942	50.5
1982	0.844	0.077	0.847	45.3	2005	0.949	-0.020	0.949	50.9
1983	0.871	0.082	0.875	46.7	2006	0.946	-0.056	0.948	50.8
1984	0.907	0.072	0.910	48.7	2007	0.943	-0.092	0.948	50.6
1985	0.934	0.039	0.934	50.1	2008	0.931	-0.137	0.941	50.0
1986	0.946	-0.004	0.946	50.8	2009	0.900	-0.171	0.916	48.3
1987	0.946	-0.038	0.947	50.8	2010	0.866	-0.180	0.884	46.5
1988	0.949	-0.077	0.952	50.9	2011	0.838	-0.177	0.857	45.0
1989	0.938	-0.123	0.946	50.3	2012	0.810	-0.171	0.828	43.5
1990	0.911	-0.161	0.925	48.9	2013	0.777	-0.146	0.790	41.7
1991	0.878	-0.174	0.895	47.1	2014	0.758	-0.100	0.765	40.7
1992	0.852	-0.179	0.870	45.7	2015	0.756	-0.054	0.758	40.6
1993	0.821	-0.178	0.840	44.0	2016	0.763	-0.014	0.763	41.0
1994	0.786	-0.159	0.802	42.2	2017	0.773	0.028	0.774	41.5
1995	0.763	-0.116	0.772	41.0	2018	0.802	0.067	0.805	43.0
1996	0.758	-0.072	0.762	40.7	2019	0.838	0.081	0.842	45.0
1997	0.759	-0.032	0.759	40.7	2020	0.870	0.079	0.874	46.7
1998	0.766	0.013	0.766	41.1	2021	0.894	0.070	0.897	48.0
1999	0.791	0.056	0.793	42.4	2022	0.926	0.052	0.927	49.7
2000	0.826	0.076	0.830	44.4	2023	0.946	0.011	0.946	50.8
2001	0.856	0.078	0.860	46.0	2024	0.953	-0.035	0.953	51.1
2002	0.883	0.077	0.887	47.4	2025	0.944	-0.074	0.947	50.7

Table 3. Inclination vector drift for a sequence of calendar years, starting each January 1 at 00:00 UTC with inclination 0°. The x- and y-components of the inclination vectors and the absolute inclination are listed for December 31 at 24:00 UTC, as well as the ΔV to move only i_x back to 0°. The rate is 53.7 m/s for each degree of inclination.

Eclipse date	Penumbra entry	exit	Sepa-rat.	Moon disc	Sun disc	Sidereal angle
1988/02/17	21:34	21:52	0.53	0.26	0.27	138.
1988/03/18	0:23	0:54	0.09	0.25	0.27	188.
1988/04/16	3:12	3:58	0.04	0.26	0.27	239.
1988/08/12	20:43	20:58	0.55	0.23	0.26	127.
1988/09/11	0:45	1:20	0.01	0.22	0.26	198.
1990/06/22	22:27	23:1	0.36	0.25	0.26	158.
1990/08/20	3:7	3:52	0.45	0.25	0.26	237.
1990/08/20	10:50	11:44	0.04	0.29	0.26	350.
1991/02/14	4:58	6:50	0.18	0.25	0.27	271.
1991/02/14	7:48	9:44	0.42	0.27	0.27	313.
1991/06/12	20:42	21:8	0.50	0.26	0.26	128.
1991/12/06	0:38	1:14	0.05	0.23	0.27	198.
1992/02/03	6:54	8:52	0.32	0.26	0.27	300.
1992/06/01	1:19	1:21	0.76	0.24	0.26	202.
1992/11/24	1:59	2:39	0.43	0.24	0.27	222.
1993/05/21	4:0	4:16	0.61	0.24	0.26	248.
1993/05/21	21:20	21:59	0.30	0.24	0.26	142.
1994/06/09	13:12	14:20	0.09	0.27	0.26	24.
1994/11/03	20:49	21:26	0.07	0.26	0.27	136.

Table 4.   The 38 partial eclipses of the Sun by the Moon seen by a
spacecraft at longitude 0° during 15 years. The times (UTC) of entry into
and exit from the penumbra is listed to the left. The next three columns
give the separation angle of the centres of the Moon and Sun and the radii
of their respective discs, in degrees, as seen from the spacecraft at the
eclipse mid-point time. The last column gives the sidereal angle (in degrees) of the spacecraft minus the sidereal angle of the Moon. Continued
on next page.

Table 4   295

Eclipse date	Penumbra entry	exit	Sepa- rat.	Moon disc	Sun disc	Sidereal angle
1995/04/29	8:50	10:7	0.08	0.27	0.26	327.
1995/05/29	12:58	13:7	0.71	0.27	0.26	14.
1995/09/24	5:5	7:26	0.35	0.27	0.27	282.
1996/03/19	2:49	3:29	0.22	0.25	0.27	230.
1997/04/07	12:42	13:32	0.53	0.31	0.27	14.
1997/04/07	19:45	20:32	0.46	0.26	0.27	116.
1998/07/23	10:18	11:20	0.44	0.29	0.26	343.
1999/01/17	9:23	10:38	0.06	0.28	0.27	331.
1999/02/16	1:25	1:59	0.19	0.24	0.27	204.
1999/02/16	14:36	16:27	0.23	0.29	0.27	44.
1999/08/11	2:31	3:10	0.12	0.25	0.26	226.
2000/01/06	7:46	9:40	0.13	0.26	0.27	314.
2000/12/25	8:43	9:21	0.58	0.27	0.27	320.
2001/01/24	20:57	21:41	0.27	0.23	0.27	132.
2001/07/20	22:35	23:3	0.46	0.25	0.26	158.
2001/11/15	14:23	16:48	0.27	0.28	0.27	52.
2002/01/13	20:57	21:40	0.30	0.23	0.27	133.
2002/05/12	12:28	13:26	0.06	0.28	0.26	14.
2003/01/02	22:46	23:21	0.40	0.24	0.27	163.

Table 4, continued. The 38 partial eclipses of the Sun by the Moon seen by a spacecraft at longitude 0° during 15 years. The times (UTC) of entry into and exit from the penumbra is listed to the left. The next three columns give the separation angle of the centres of the Moon and Sun and the radii of their respective discs, in degrees, as seen from the spacecraft at the eclipse mid-point time. The last column gives the sidereal angle (in degrees) of the spacecraft minus the sidereal angle of the Moon.

LIST OF SYMBOLS

The list contains symbols that are used in more than one section. The number in parenthesis refers to the section where the symbol first appears or is defined.

A = Unperturbed geostationary semimajor axis (2.3) = 42164.2 km

a = Orbit semimajor axis (2.2) $\approx$ 42165.8 km

B = Tangential acceleration from Earth's gravity field (4.2)

b = 1st coefficient in time-dependence of tracking (8.5)

c = speed of light (8.2) = 299792.458 km/s

D = Mean longitude drift rate (2.3)

$\bar{e}$ = (e_x, e_y) = Eccentricity vector (2.3)

e = Orbit eccentricity (2.2)

$\bar{F}$ = Force vector (2.1, 3.1)

f = Earth's oblateness (8.1) = 1/298.257

G = Greenwich sidereal angle (2.1)

$\bar{I}$ = 3-dimensional inclination vector (2.2)

$\bar{i}$ = (i_x, i_y) = 2-dimensional inclination vector (2.3)

i = Orbit inclination (2.2)

j = Imaginary unit (4.2, 6.4, 8.6, 8.7)

l = Ground station longitude (8.1)

m = Spacecraft mass (1.2, 2.1)

P = Solar radiation pressure (4.5) = 4.56×10^{-6} N/m^2

p = Solve-for parameter in orbit determination (8.4)

Q = Matrix of partial derivatives of tracking measurements (8.4)

q = Tracking measurement of any type (8.4)

R = Earth's radius at equator (1.2, 8.1) = 6378.144 km

$\bar{R}$ = Ground station vector in Earth-rotating system (8.1)

r = Distance from Earth's centre to spacecraft (1.2) $\approx$ 42164.5 km

$\bar{r}$ = (x, y, z) = Vector from Earth's centre to spacecraft (2.1)

s = Sidereal angle of spacecraft (2.1)

t = Time, as independent variable (2.1)

u = 2nd coefficient in time-dependence of tracking (8.5)
V = Spacecraft velocity (1.2) $\approx$ 3.075 m/s
$\overline{V}$ = Spacecraft velocity vector (2.2)
v = 3rd coefficient in time-dependence of tracking (8.5)
W = Weighting matrix (8.4)
w = 4th coefficient in time-dependence of tracking (8.5)
α = Azimuth angle from station to spacecraft (8.1)
β = Auxiliary angle in station - spacecraft geometry (8.3)
γ = Auxiliary angle in station - spacecraft geometry (8.3)
Δ = Instantaneous change of a variable by a manoeuvre (3.1)
δ = Difference of a variable, not by manoeuvres
ε = Error in an orbit parameter (2.3)
ε = Elevation angle from station to spacecraft (8.1)
θ = Spacecraft declination = spacecraft latitude (2.1)
ϑ = Angle between spacecraft and station from Earth's centre (8.1)
λ = Spacecraft longitude (2.1)
μ = Gravity potential of Earth (1.2) = 398600.440 km^3/s^2
v = Orbit true anomaly (2.2)
Ξ = Matrix for least squares fit of tracking (8.6)
ρ = Range = distance from ground station to spacecraft (8.1)
$\overline{\rho}$ = Position vector of spacecraft in topocentric system (8.1)
σ = Effective cross-section-to-mass ratio (4.5)
σ = Error standard deviation (6.5, 8.6)
Φ = Matrix of partial derivatives (8.5)
ϕ = Ground station latitude (8.1)
Ψ = Matrix of partial derivatives (8.5)
ψ = Earth's rotation rate (1.2) = 0.729211585 $\times$ 10^{-4} rad/s
Ω = Right ascension of orbit ascending node (2.2)
ω = Argument of perigee of orbit (2.2)

BIBLIOGRAPHY

The Astronomical Ephemeris. Issued every year by her Majesty's Stationary Office, London.

Explanatory Supplement to the Astr. Ephem. and the American Ephemeris and Nautical Almanac. Her Maj.'s Stationary Office, London 1974

Final Acts. Adopted by the Second Session of the World Administrative Radio Conference on the Use of the Geostationary-Satellite Orbit and the Planning of Space Services Utilizing It (ORB-88). International Telecommunication Union, Geneva 1988

Proceedings of a CNES Colloquium in October 1985 on: Space Dynamics for Geostationary Satellites - Mecanique Spatiale pour les Satellites Geostationaire. Cepadues-editions, Toulouse 1986

Agrawal B.N.: Design of Geosynchronous Spacecraft. Prentice Hall 1986

Brumberg V.A.: Essential Relativistic Celestial mechanics. Hilger Ltd, Bristol 1991

Chobotov V.A. (Ed.): Orbital Mechanics. AIAA Education Series, Washington 1991

Christol C.Q.: The modern International Law of Outer Space. Pergamon Press, New York 1982 & 1984

Elyasberg P.E.: Introduction to the Theory of Flight of Artificial Satellites. Israel Program for Scientific Translations 1967

Escobal P.R.: Methods of Orbit Determination. Wiley, New York 1976

Gaposchkin E.M., Kolaczek B. (Ed.): Reference Coordinate Systems for Earth Dynamics. Reidel Publishing Company, Dordrecht 1981

Garner J.T., Jones M.: Satellite Operations. Systems approach to design and control. Libr. Space Sci. Space Techn., Chichester 1990

Lambeck K.: The Earth's Variable Rotation. Cambridge University Press 1980

Larson W.J. (Ed.): Space Mission Analysis and Design; 2nd Ed. Kluwer Academic Publishers, Dordrecht 1992

Neutsch W., Scherer K: Celestial Mechanics. An Introduction to Classical and Contemporary Methods. Wissenschaftsverlag, Mannheim 1992

Pocha J.J.: An Introduction to Mission Design for Geostationary Satellites. Reidel Publishing Company, Dordrecht 1987

Prussing J.E., Conway, B.A.: Orbital Mechanics. Oxford University Press 1993

Rimrott F.P.J.: Introductory Orbit Dynamics. Vieweg & Sohn Braunschweig/Wiesbaden 1989

Rosser J.B. (Ed.): Space Mathematics Part 1. Lectures in Applied Mathematics, American Mathematical Soc., Providence 1966

Roy A.E.: Orbital motion. Hilger Ltd, Bristol 1978

Seidelmann P.K.: Explanatory Supplement to the Astronomical Almanac. University Science Books 1992

Smith M.L.: International Regulation of Satellite Communication. Utrecht Studies in Air and Space Law 7. Kluwer Academic Publishers, Dordrecht 1990

VanFlandern T.C., Pulkkinen K.F.: Low-Precision Formulae for Planetary Positions. Astroph. J. Suppl. Ser. 41, 391-411, Nov. 1979

Wertz J.R., Larson W.J. (Ed.): Space Mission Analysis and Design. Kluwer Academic Publishers, Dordrecht 1991

SUBJECT INDEX
(Numbers refer to sections)